AF711533

Cover: *Soldiers of the 16th Infantry around a campfire at San Gerónimo, Mexico, May 1916* (Library of Congress)

CMH Pub 77–1

THE MEXICAN EXPEDITION 1916–1917

by
Julie Irene Prieto

Reprinted by St. John's Press, Alexandria, Va.
978-1-944961-45-9
2016
www.stjohnspresspublishing.com

Center of Military History
United States Army
Washington, D.C., 2016

INTRODUCTION

A century ago, the United States joined what was then called the "Great War." The conflict, and America's involvement that helped bring it to an end, cast a long shadow across the twentieth century. In addition to massive destruction and loss of life, it shattered the European power system and brought economic turmoil, revolutions, collapse of empires, the birth of new nation-states, and the rise of totalitarian movements.

The war's outbreak in August 1914 shocked the United States. At this time, the Army was a constabulary force, barely adequate to police America's new empire in the Caribbean and Pacific, although after the War with Spain in 1898, Secretary of War Elihu Root had led important reforms. The United States invested in coastal defenses and a fleet to guard its shores, but its Army was simply too small to make an impact on European battlefields and lacked modern equipment such as motorized vehicles, machine guns, rapid-firing artillery, airplanes, and tanks. Its largest standing formation was a regiment. Not until June 1916 did Congress authorize an expansion of the Army, dual state-federal status for the National Guard, and the creation of an Army Reserve.

Meanwhile, the United States faced threats closer to home. To the south, Mexico was in the throes of revolution. In March 1916, a cross-border raid on Columbus, New Mexico, caused President Woodrow Wilson to declare a partial mobilization for a "punitive expedition" and to defend the border.

In early 1917, the president decided that America had run out of diplomatic options. In April, he asked Congress to declare war against Germany. But the U.S. Army, numbering only 133,000 men, was far from ready. The president ordered nearly 400,000 National Guardsmen into federal service, while the War Department built thirty-two cantonments where new divisions could be mobilized and trained. On 5 June, 9.5 million men registered for America's first conscription since the Civil War. The Army created a new combined-arms formation—the square division—and then corps, field armies, and support units, as well as new branches such as the Air Service, the Tank Corps, and the

Chemical Warfare Service. By the end of 1918, the Army grew to 3.7 million men and trained 200,000 new officers to lead them.

In June 1917, the 1st Expeditionary Division deployed to France, arriving in time to parade through Paris on the Fourth of July. On that occasion, a spokesman for Maj. Gen. John J. Pershing declared at the tomb of the hero of the American Revolution, "Lafayette, we are here!" In September, the first National Guard division deployed, the 26th Division from New England. By war's end, the American Expeditionary Forces had grown to two million soldiers and more than forty divisions. Among them were over 200,000 African American soldiers who served in segregated units, including four infantry regiments that fought under French command.

During 1918, American "doughboys" learned to fight in battles of steadily increasing scale: Cantigny, the Marne, Aisne-Marne, St. Mihiel, and Meuse-Argonne, adding thirteen campaign streamers to the Army flag. The soldiers faced machine guns, tanks, and a frightening new weapon of mass destruction, poison gas. Overall, in little over six months of combat, the American Expeditionary Forces suffered more than 320,000 casualties, including 50,280 killed.

The war that had promised to "make the world safe for democracy" ended with an armistice on 11 November 1918, followed by a deeply flawed peace. Although the U.S. Senate refused to ratify the Treaty of Versailles, American soldiers served in the Occupation of the Rhineland until 1923, before withdrawing from Europe altogether. The war sowed the seeds of future conflict that within a generation led to a second and even greater world war.

The United States will never forget the American soldiers who fought and died in the First World War. America's first unknown soldier was laid to rest on 11 November 1921 in the Tomb of the Unknowns in Arlington National Cemetery, where soldiers still stand guard. The United States created permanent American military cemeteries in France, Belgium, and Britain to bury the fallen. To this day, memorials to their sacrifice can be found across America. The last surviving U.S. Army veteran of the war did not die until 2011. It is to all the doughboys, those who returned and those who did not, that the U.S. Army Center of Military History dedicates these commemorative pamphlets.

JAMES C. MCNAUGHTON
Director, Histories Division

THE MEXICAN EXPEDITION 1916–1917

On 9 March 1916, the forces of Doroteo Arango, better known as Francisco "Pancho" Villa, attacked the small border town of Columbus, New Mexico. In response to the raid, President Woodrow Wilson authorized Brig. Gen. John J. "Black Jack" Pershing to organize an expedition into Chihuahua, Mexico, in order to kill or capture Villa and those responsible for the assault. By 15 March, 4,800 Regular Army soldiers had assembled in Columbus and Camp Furlong, the Army garrison just outside of the town's center. These men fanned out into the Mexican countryside on horseback in small, highly mobile cavalry detachments—sometimes led by local guides or by the Army's Apache scouts—that could cover large swaths of sparsely populated and rough terrain. Cavalrymen employed skills and strategies developed in the preceding decades on frontier campaigns in the West and in warfare against irregular, guerrilla forces in the Philippines. The Mexican Expedition, popularly called the "Punitive Expedition," was to be one of the last operations to employ these methods of warfare and one of the first to rely extensively on trucks. It also provided a testing ground for another new technology—the airplane.

During the eleven months that Pershing's expedition was in Chihuahua, U.S. troops failed to kill, capture, or even spot Pancho Villa, but the impact of the expedition reached far beyond the deserts of northern Mexico. The approximately 10,000 regulars that served in the Punitive Expedition gained experience in large, multiunit field operations at a time when small-unit actions were the norm. At the same time, the task tested Pershing as a commander, leader, and organizer. Although he did not always agree with Secretary of War Newton D. Baker and President Wilson, Pershing proved that he could run an efficient campaign despite the difficulties of operating in northern Mexico and the restrictions that Washington placed on his operations. In addition, although the approximately 110,000 members of the National Guard who patrolled the border from May 1916 through February 1917 saw little action, they received training that bolstered Army readiness just months before the United States declared war on Germany

Pancho Villa, military leader of rebel forces during the Mexican Revolution and considered a bandit by Americans in the wake of the raid on Columbus, New Mexico (Library of Congress)

in April 1917. Thus, the U.S. Army was far more prepared for war in Europe than it had been in early 1916. Despite this unexpected benefit to the Army, the results of the Punitive Expedition were not all positive for the United States. Wilson's invasion of Mexican territory and intervention in the Mexican Revolution created an environment of suspicion and distrust that took decades to repair and caused a general decline in relations between the United States and Latin American republics.

Strategic Setting

From 1910 to 1920, revolution gripped Mexico. Since 1876, dictator Porfirio Díaz had ruled Mexico with only one interruption. In an attempt to spur economic development, Díaz offered incentives and enacted laws that allowed foreign individuals and corporations to dominate large sectors of the nation's economy. Although this stimulated some growth, foreign entities came to own more than 100 million acres of territory while 90 percent of Mexican peasant farmers owned none. By 1910, U.S. interests held

97 percent of mining concerns and controlled 90 percent of the petroleum industry in Mexico. Though dictatorial, Díaz maintained a façade of elections and constitutional rule throughout the length of his regime.

In 1908, Díaz announced he would not seek re-election in 1910, leading political moderate Francisco I. Madero to campaign aggressively for the presidency. After realizing his rival might win office, Díaz reversed his position on retirement and arrested Madero, who escaped jail into exile. From San Antonio, Texas, Madero issued a revolutionary manifesto, the Plan de San Luis Potosí, which called for a national uprising against the dictatorship. A brief, popular revolt followed, after which Madero became Mexico's new president in 1911.

Long-term disaffection with Díaz by the Mexican people, coupled with the events surrounding the election of 1910, sparked a protracted series of conflicts collectively known as the Mexican Revolution. Concerned about the lives, property, and investments of U.S. citizens living in Mexico and the border region, the War Department sent Army units to the border on several occasions during the war. The first major deployment occurred in 1911 when President William H. Taft ordered 20,000 soldiers to Galveston and San Antonio, Texas, and San Diego, California. Officially designated as the "Maneuver Division" for what the War Department described as training exercises, or "war games," it served to warn Mexican authorities that the United States could intervene militarily in the conflict to protect its citizens and their interests. While the Maneuver Division enjoyed widespread support when it first was formed, Secretary of State Philander C. Knox came to distrust the measure and expressed fears that moving troops to the border might encourage Washington to use invasion before diplomacy as a tool for enacting change, resulting in a long and costly occupation south of the border. In any case, the United States began withdrawing its forces as Madero consolidated his hold on power, and the Maneuver Division disbanded on 7 August 1911.

Secretary Knox was correct that occupying northern Mexico—particularly the state of Chihuahua where the Punitive Expedition ultimately operated—would have been challenging due in part to its difficult terrain. Chihuahua, Mexico's largest state, roughly the size of Michigan, consists of a high plateau that rises into the Sierra Madre Occidental with elevations of up to 10,000 feet. Although the state has a desert climate, the high elevation creates extremes in temperature, with blazing hot days in the summer and frigid blizzards in winter and spring. Temperatures can rise or fall thirty or forty degrees within the same day, and sand quickly gives way to grasslands and forest depending on the elevation. In the early twentieth century, this remote area had few roads or telegraph lines to link the small towns and ranches that dotted the state. Except for connections with the state capital, Chihuahua city, telegraph lines did not extend much farther south than the town of Namiquipa, about 130 miles south of the border, presenting U.S. forces with a major challenge in maintaining basic communication. Large-scale cattle ranches, including the one million-acre farm at Babícora owned by U.S. newspaper tycoon William Randolph Hearst, as well as mining operations divided much of the land. The state also fell victim to violent dust storms that periodically coated entire regions with a fine layer of silt. Two major railroads, the Mexican National and the Northwestern, ran north-south, but only a few spur lines covered the rest of the state.

The border between the United States and Mexico remained calm again until February 1913 when Madero was assassinated during a coup led by General Victoriano Huerta. This touched off years of renewed fighting that pitted Huertista forces against rebels led by Venustiano Carranza, a shrewd, pragmatic, Coahuilan from a well-to-do ranching family, whose supporters were known as Constitutionalists. Pancho Villa stood as one of Carranza's allies. Before the coup, Villa had backed Madero until the president decided to arrest him on Huerta's recommendation that he neutralize possible rivals. After escaping from prison in Mexico City prior to Madero's death, Villa fled to the United States and lived for a short time in exile until the coup, when he crossed the border again to fight against Huerta. Once in Mexico, Villa raised his own army, a well-organized force numbering 30,000 to 50,000 men at its peak, that became known as the *División del Norte* after its geographical area of operation. The charismatic Villa led his men to a series of victories on the battlefield, compensating for a lack of arms, ammunition, and supplies with unpredictable tactics, such as night attacks that surprised and terrorized his enemies. As Villa's army rose in prestige in northern Mexico, the sometime-outlaw-turned-populist-revolutionary general grew in lore. In the United States, he became a well-known, almost romantic figure and even made the acquaintance of Pershing and Army Chief of Staff Maj. Gen. Hugh L. Scott in 1914. Pershing saw him as a potential ally who could protect U.S. interests as revolution consumed Mexico, and he gave Villa texts and studies on the art of war.

As this renewed wave of violence deepened, President Wilson came to believe that the Constitutionalists would be unable to remove Huerta from power without aid from the United States. Even though Carranza was opposed to U.S. military intervention in Mexico, Wilson sought a reason to intervene in the struggle between these two powers. The president finally seized upon a minor incident that occurred on 9 April 1914 in which nine U.S. sailors were briefly arrested in a Tampico fueling station after straying into an off-limits area. Mexican officials quickly released the sailors and apologized but refused to perform a twenty-one gun salute in penance, as requested by R. Adm. Frank Friday Fletcher, head of the naval forces there. In response, Wilson asked Congress for authorization to occupy the port of Veracruz, about 280 miles south of Tampico. While Wilson waited for Congress to act, the president received a report that a ship, the *Ypiranga*, approached Veracruz with weapons destined for Huerta. Wilson ordered the Navy to seize the customs house at Veracruz and prevent the arms from reaching their destination. Marines and sailors landed on 21 April, followed ten days later by a brigade of soldiers. The combined force eventually occupied the entire city of Veracruz. Fletcher later received the Medal of Honor for leading the successful landing in Mexico.

Carranza never accepted the presence of U.S. troops. Since 1898, the United States had invaded a string of Latin American countries. At the time of the seizure of Veracruz, U.S. forces already occupied Nicaragua, Haiti, and the Dominican Republic. However tempting Wilson's support might have been, Carranza did not want to risk Mexico becoming a protectorate of its neighbor to the north. Even with his unwillingness to accept U.S. aid, Carranza's Constitutionalists succeeded in ousting Huerta on 15 July 1914. Brig. Gen. Frederick N. Funston administered Veracruz city until 23 November when U.S. forces withdrew. By that time, World War I had erupted in Europe, capturing the attention of U.S. citizens.

Soon after Huerta fled into exile, Villa split with Carranza's regime and joined with Emiliano Zapata, a revolutionary leader who operated mainly in his native Morelos, south of Mexico City, against the Constitutionalists. Villa's and Zapata's combined forces attacked and briefly held Mexico City in April 1915, but that summer Villa suffered a significant defeat against Carranza's armies at Celaya. Villa lost more battles as well as influence through summer, and by September mounting casualties reduced the once formidable *División del Norte* to a small guerrilla band of between 500 and 1,000 men. Many of these men had formed part of Villa's elite guard, the *Dorados*. Young and loyal to their general, they mostly hailed from either Chihuahua or Durango. Some had been popular leaders in their hometowns and could recruit and lead small bands of men through personal sway and act semi-independently in the field. Due in part to the U.S. arms embargo on Villa, these men had few weapons and lacked adequate ammunition to conduct regular warfare, opting instead for quick raids on vulnerable and remote positions followed by retreat.

For Villa, President Wilson's decision to officially recognize the intelligent but uncharismatic Carranza as the legal president of Mexico proved the final insult. This move allowed the Constitutionalist leader to legally import arms from the United States to use against Villa and other opposing revolutionary forces. In the wake of this decision, Villa, embittered by the loss of U.S. support, lashed out at his former allies and became convinced that a conspiracy was afoot. He came to believe that Wilson and Carranza had entered into a corrupt bargain to sell northern Mexico to the United States in exchange for political recognition. Acting from his home base in Chihuahua and Durango, Villa charged Carranza with treason and railed against Wilson for violating U.S. neutrality.

With his army just a shadow of its former existence, Villa adopted new tactics to continue his fight against Carranza's better-equipped forces. In Chihuahua, the general turned to impressment for the first time to fill his depleted ranks. Villa also began raiding Chihuahua's small and increasingly disaffected middle class for supplies. This represented a marked change, since through 1915, Villa had managed to provision his army mostly by confiscating goods and money from the wealthiest hacendados in the state. Where Villa once had avoided attacking U.S. citizens living and working in Mexico, his forces now targeted them in retaliation for Wilson siding with Carranza. The deadliest result of this new policy came on 10 January 1916 near Santa Isabel, Chihuahua. There, in the desert, a detachment of Villistas halted and boarded a train that had eighteen American engineers on board. Employed by the Cusihuiriachic, or "Cusi," Mining Company, they were on their way to reopen the mines and carried passes of safe conduct issued by Carranza's government. Villa's men ordered the Americans off the train and shot them execution-style, save one who escaped (*Map 1*).

In response, Carranza's forces prepared a new offensive to stamp out Villa's remaining troops. The main government force in

Chihuahua was led by General Jacinto Treviño, who commanded the 1st, 3d, and 5th Divisions of the Army of the Northeast. These three divisions consisted of approximately 10,231 troops, led by 6 subordinate generals and 41 other officers. Unfortunately, Treviño and many of his troops were from Coahuila, not Chihuahua. To the local population, having forces from outside of the state operating within its borders was akin to being occupied by a foreign force. Local military leaders and troops alike often refused to follow Treviño's commands, showcasing how weak support for Carranza's regime was in some areas far from the capital. Local antagonism toward Treviño only worsened after 12 May 1916, when Carranza recalled Governor Ignacio C. Enríquez of Chihuahua, probably at Treviño's insistence. Carranza then replaced Enríquez with the general's brother, Francisco Treviño, who proceeded to use his office for personal enrichment. Beyond the friction that festered between Treviño, local military commanders, and the citizens of Chihuahua, Carrancista forces suffered from a host of problems including a dearth of discipline, high rates of desertion, and a dangerous lack of weapons and other supplies.

With violence on the border rising, Col. Herbert J. Slocum, commander of the 13th Cavalry at Camp Furlong, tried to determine where and when Villa might attack next. The 21 officers and 532 soldiers of his regiment patrolled sixty-five miles of border, a difficult task made even more complicated by conflicting reports of Villa's movements. On 7 March, Slocum spoke to two Mexican men who claimed to have seen Villa's camp about forty miles south of the border. The colonel offered one of the men, Antonio Muñoz, $20 for additional information. Although clearly reluctant, Muñoz crossed into Mexico and returned the next day, reporting to Slocum that he had seen about 120 Villistas at Boca Grande, thirty miles south of the border. Muñoz claimed that they seemed to be moving south toward Parral. On the same day, Maj. Gen. Frederick N. Funston—in command of the Army's Southern Department, encompassing Texas, Oklahoma, New Mexico, and Arizona—alerted Slocum that the Department of State had received a report that Villa might go to Columbus, about eighty miles west of El Paso, Texas. They reported that Villa might either ask for asylum or cause an incident to provoke U.S. intervention in Mexico. Pershing, the commander of the Army's 8th Brigade at Fort Bliss, also received a warning from Mexico's General Gabriel Gavira that a Villista force was seen moving north toward the border. Slocum dispatched additional

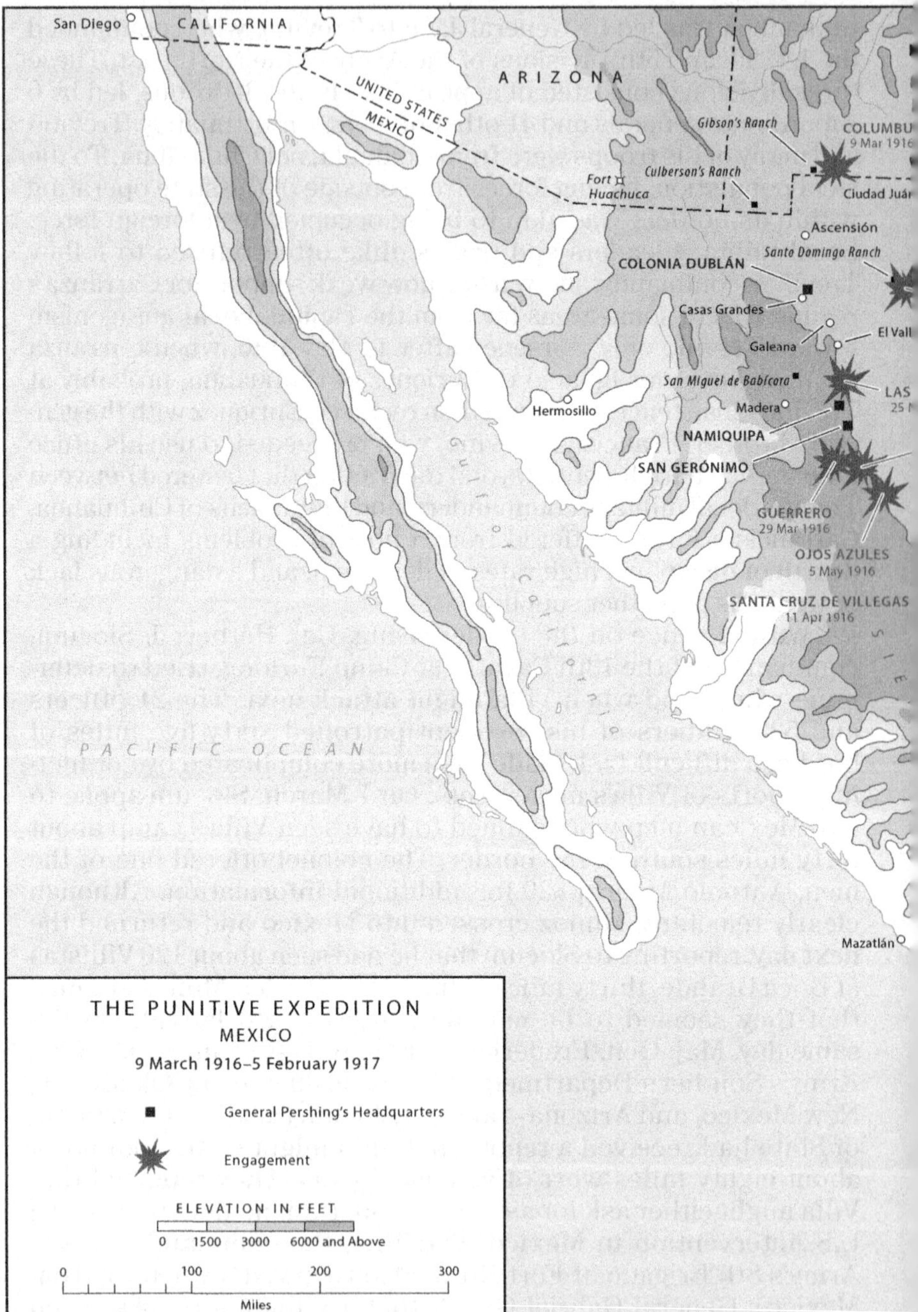

San Diego
CALIFORNIA
ARIZONA
UNITED STATES
MEXICO
Gibson's Ranch
COLUMBU
9 Mar 1916
Culberson's Ranch
Fort Huachuca
Ciudad Juár
Ascensión
Santo Domingo Ranch
COLONIA DUBLÁN
Casas Grandes
Galeana
El Vall
San Miguel de Babícora
Hermosillo
Madera
NAMIQUIPA
SAN GERÓNIMO
GUERRERO
29 Mar 1916
OJOS AZULES
5 May 1916
SANTA CRUZ DE VILLEGAS
11 Apr 1916
PACIFIC OCEAN
Mazatlán
THE PUNITIVE EXPEDITION
MEXICO
9 March 1916–5 February 1917
General Pershing's Headquarters
Engagement
ELEVATION IN FEET
0 1500 3000 6000 and Above
0 100 200 300
Miles

Map 1

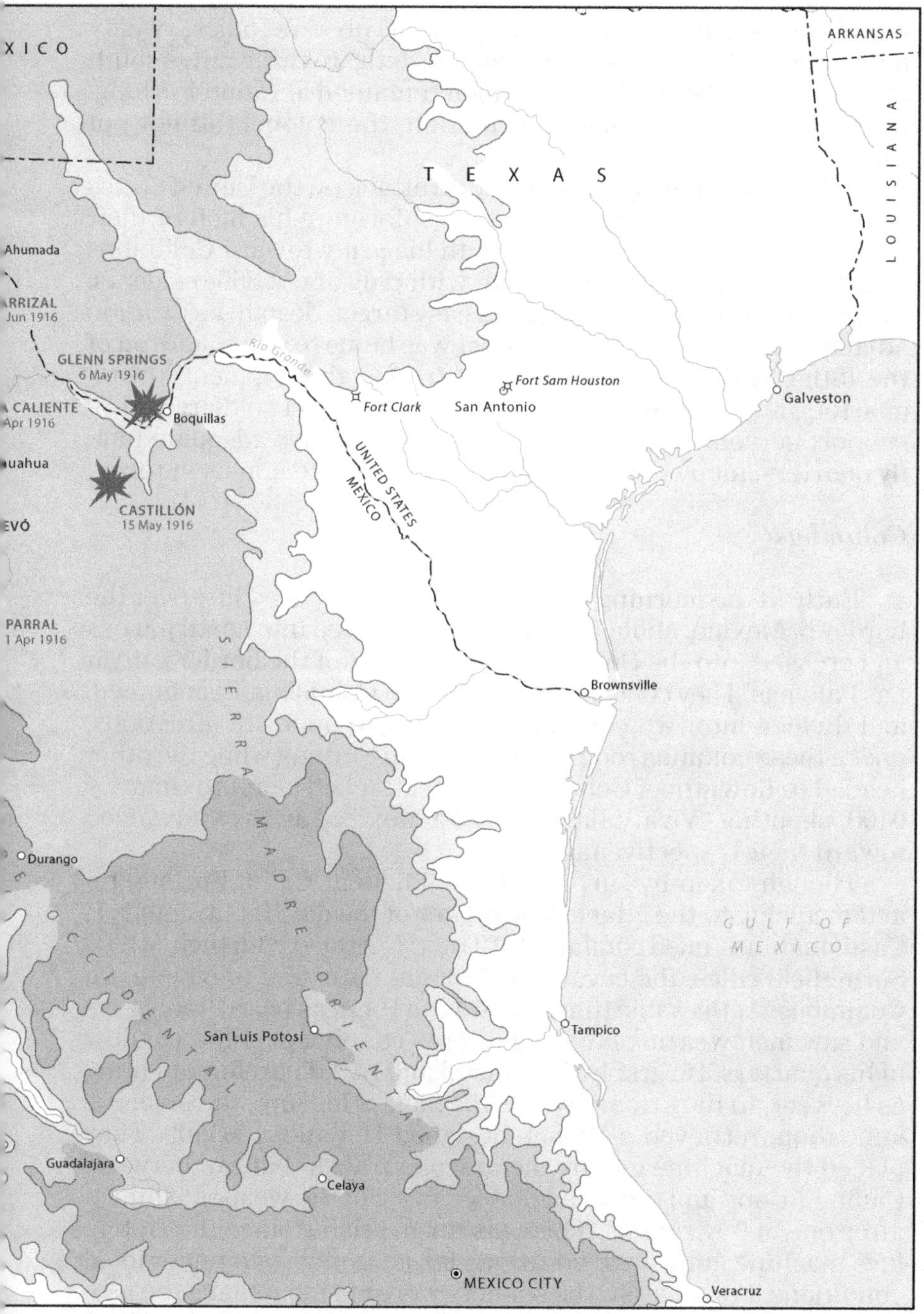

XICO
ARKANSAS
LOUISIANA
TEXAS
Ahumada
RRIZAL
Jun 1916
GLENN SPRINGS
6 May 1916
CALIENTE
Apr 1916
Boquillas
uahua
CASTILLÓN
15 May 1916
EVÓ
PARRAL
1 Apr 1916
Rio Grande
Fort Clark
Fort Sam Houston
San Antonio
Galveston
UNITED STATES
MEXICO
Brownsville
SIERRA MADRE ORIENTAL
OCCIDENTAL
Durango
GULF OF MEXICO
Tampico
San Luis Potosí
Guadalajara
Celaya
MEXICO CITY
Veracruz

patrols and strengthened the garrisons at Gibson's Ranch, twenty miles west of Columbus, and at the border gate three miles south of town. Only about 120 cavalrymen remained at Camp Furlong. In the absence of reliable information, the colonel did not put the post on alert.

In fact, Villa had been planning an attack on the United States since at least late February. Without informing his men of their objective, Villa began marching with his army toward Columbus on 24 February. Small and remote with only about 350 residents, the town presented a tempting and easy target, despite its location adjacent to Camp Furlong. The post was home to one squadron of the 13th Cavalry, with four lettered troops, the regiment's headquarters, and machine gun troops—a total of 274 soldiers and 79 support personnel. Because the post did not have adequate family quarters, many of the officers lived in private homes in town.

Columbus

Early in the morning on 9 March, as Villa waited just over the border in Mexico, about 500 Villistas organized into small parties and crossed into the United States just west of the border gate at Las Palomas. They concentrated outside of Columbus, dismounted, and divided into two columns. Traveling in opposite directions, one of these columns rode to attack Camp Furlong while the other headed to downtown Columbus. Both attacked at approximately 0400, shouting "Viva Villa" and "Viva México" as they advanced toward their respective targets.

Though taken by surprise, the soldiers at Camp Furlong reacted quickly to the attack. The officer of the day, 1st Lt. James P. Castleman, assumed command of Troop F. Armed with their M1903 Springfield rifles, the cavalrymen fought their way to downtown Columbus. At the same time, 2d Lt. John P. Lucas heard hoof beats and saw men wearing large sombreros running past the window of his quarters. He grabbed his pistol and raced barefoot, fighting as he went, to the armory where the men of his unit, the machine gun troop, retrieved a French-designed Hotchkiss M1909. They placed the machine gun by the railroad tracks that ran between Camp Furlong and town and opened fire, but the weapon jammed after only a few rounds. This was unsurprising since the Hotchkiss machine guns were notorious for jamming even under ideal conditions. Fortunately, the soldiers were able to replace the useless gun with four others from the armory. About the same time,

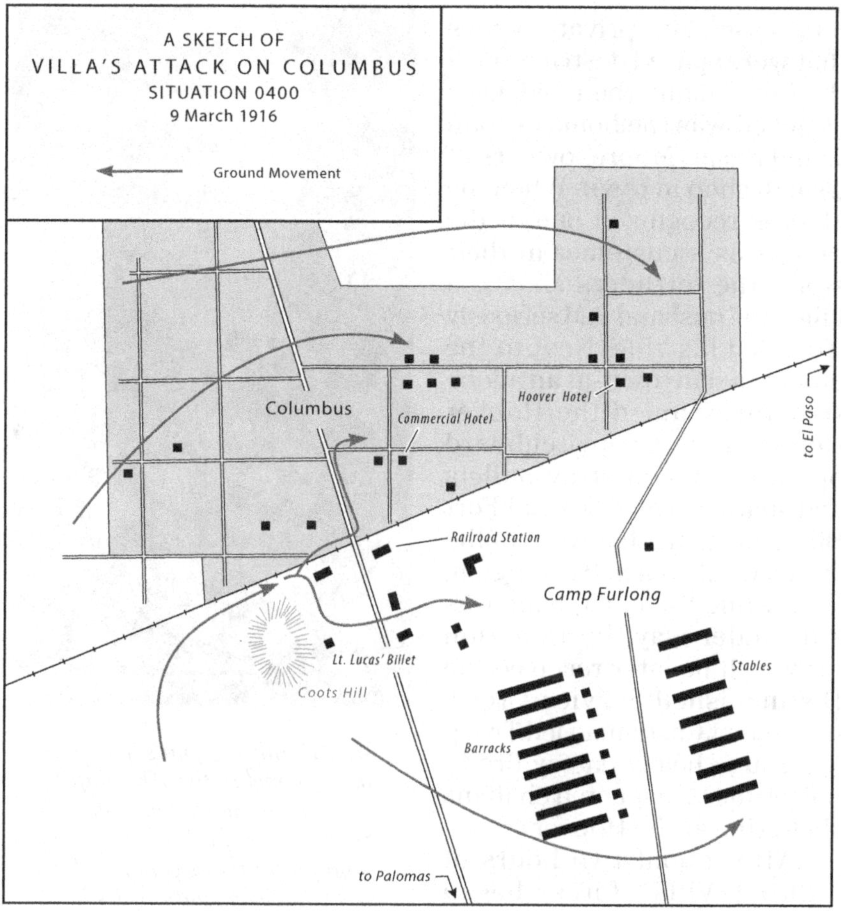

Map 2

a group of Villistas burst into the adobe building that housed the post kitchen. The cooks fought back with the shotguns they normally used for hunting along with axes, pots of boiling water, and kitchen utensils (*Map 2*).

The Villista column that advanced on downtown Columbus managed to do more damage. One small group detached from the main force and went to the Commercial Hotel, where they killed four townspeople and then robbed and set fire to the grocery store across the street. The flames from the fire spread to the Commercial Hotel and other buildings. Raiders systematically plundered the businesses and private homes that were spared destruction in the fire. Among the residences attacked was the home of John J. and Susan Moore, owners of a small shop in town. When the Moores recognized one of the raiders as a customer in their store, the intruders shot and killed the husband and seriously wounded his wife. Next to the other hotel in town, at an adobe structure named the Hoover Hotel, the town switchboard operator braved stray bullets and shattered glass to dial Fort Bliss, ensuring that word of the attack would reach Washington, D.C., while the emergency was still under way. In an action for which he later received the Distinguished Service Cross, Lieutenant Castleman led Troop F through heavy enemy fire to relieve the beleaguered civilians sheltering at the Hoover.

Maj. Frank Tompkins, pictured as a colonel in the 1930s after receiving the Distinguished Service Cross for his leadership and gallantry as commander of the 3d Squadron, 13th Cavalry, during the 1916 raid on Columbus, New Mexico (National Archives)

After about two hours of fighting, Villa's forces began retreating in the direction of the border. Slocum gave Maj. Frank Tompkins—commander of the 3d Squadron and acting regimental executive officer—permission to pursue the raiders with Troops F and H. A Villista rear guard attempted to delay the pursuit from a piece of high ground about 300 yards south of the border, but it broke and ran when Tompkins led the troopers in a mounted charge. The cavalrymen then dismounted with their rifles and engaged the fleeing Villistas, killing several before the raiders moved out of range. At this point, Tompkins realized that they had skirmished inside Mexico and thus had violated international law. He sent a messenger to Slocum to ask what to do. Slocum replied that

Tompkins should use his own judgment in making a decision, so the cavalrymen remounted and continued the pursuit into Mexico. Although outnumbered, the troopers drove the rear guard back on the enemy's main body in a series of four skirmishes and recovered some of the goods Villistas had taken at Columbus. About fifteen miles south of the border, the U.S. soldiers dismounted and repelled a counterattack. Carrying two wounded men, lacking rations, and running low on water and ammunition, Tompkins finally led the tired troopers and their exhausted mounts back to Camp Furlong. For this action, Tompkins later received both the Distinguished Service Cross and the Distinguished Service Medal.

Even though the Villistas retreated into the desert, the violence in Columbus did not end with Tompkins' return. In the immediate aftermath of the raid, rumors spread that the Villistas had selectively targeted Anglo American businesses and citizens at Columbus, leading to a panic that Mexican Americans in town had acted as spies for the Villistas. Even though he had no evidence to support these charges, Slocum temporarily declared martial law and ordered members of the 13th Cavalry to search several area homes. An unknown number of people were detained on charges of espionage or theft and forced to leave town. Over the course of the next several days, reports circulated of extralegal killings of up to twelve Mexicans and Mexican Americans.

When viewed as a military operation, the raid on Columbus was a failure. Although they captured about 80 horses, 30 mules, and 300 rifles and shotguns, Villa's army sustained significant losses, including about 100 dead. Sixty-seven of these fell in and around Columbus, while the rest perished in the retreat. Thirty more Villistas were captured, and an unknown number were wounded. Some of these losses were the result of poor planning and intelligence. Prior to the raid, scouts had reported to Villa that only about 50 soldiers were stationed at Camp Furlong when in fact more than 300 manned the post. Villa's troops still had the numerical advantage in Columbus, but they were poorly armed and dangerously short on ammunition. In order to attack swiftly, devastate the town, and withdraw, Villistas would have needed to sweep through the town relatively unopposed, but Camp Furlong had too many U.S. troops to make this possible. In contrast, the United States suffered fewer losses, with eight soldiers killed and five wounded. Ten civilians also were killed

Trooper of the 13th Cavalry on guard against looting following Villa's raid on Columbus, New Mexico (Library of Congress)

in town. Authorities learned later that just before the attack, while en route to Columbus, Villistas hanged two U.S. citizens and shot a third ranch hand they had taken captive in Mexico.

The Villista raid on Columbus may have failed as a military action, but it constituted a startling political and strategic victory for the rebel leader. Villa stood in a difficult position in the first months of 1916, and he knew that in his weakened state, he needed to do something that would destabilize Carranza's regime. One way to do this was to prompt an invasion from the United States, and launching an attack within U.S. territorial boundaries would achieve this end. Even if he had wanted to minimize the attack, Wilson felt tremendous pressure to act due to the upcoming presidential election, and the president hardly could ignore the extensive press coverage of the raid and the genuine popular outrage it generated. Villa calculated that military intervention from the north would prompt one of two possible outcomes. Either Carranza would oppose the presence of U.S. troops, leading to a diplomatic break between Mexico and the United States, or Carranza would support U.S. actions in which case patriotic Mexican citizens would turn against the regime. Believing in Carranza's "corrupt bargain" with Wilson, Villa also may have seen attacking the United States as a way to exact revenge over his former ally for what he saw as a betrayal of Mexican sovereignty.

As Villa expected, the United States did move quickly to send troops to Mexico in the wake of the raid on Columbus. On 10 March,

General Funston sent Wilson a telegram recommending that he authorize an expedition to "pursue into Mexican territory hostile Mexican bandits who raid American territory." Funston argued in this communication that the inadequately policed border allowed Villa to "harass our ranches and towns" with impunity and that sending in an expedition would remedy the situation. After meeting with Secretary of War Newton D. Baker and General Scott, the president accepted Funston's proposal. The Southern Department commander expected that he personally would lead the expedition, but on Scott's recommendation, the president appointed Pershing. While originally focused only on apprehending Villa, Wilson ordered Pershing to capture or disband the Villista units that had attacked Columbus and then withdraw from Mexico as soon as possible.

In the spring of 1916, gathering the required number of Army regulars to man what was termed the "Punitive Expedition" was a major undertaking. Part of the reason for this was that the U.S. Army was relatively small, consisting of only 108,399 soldiers. The Army stationed most of these soldiers in the U.S.-administered Philippines, Hawaii, and the Panama Canal Zone, while 24,602 others were scattered across the continental United States. Many of these men had experience with colonial operations in the Philippines, policing the islands, or fighting the small guerrilla forces dispersed throughout the countryside.

In contrast, European armies expanded rapidly after 1914 through the use of conscription. Great Britain's army, comprised of just 247,432 regulars in 1910, swelled to nearly 4 million. France had averaged 670,000 active troops before the war but called up 2.9 million for active service when the war broke out in 1914. By the end of the war, approximately 8,317,000 Frenchmen had served. Germany had a prewar force of about 800,000 men, but roughly 13,250,000 had served by 1918. The soldiers conscripted in Europe fought as members of large, multiunit forces in battles that pitted millions of men against each other for months at a time and required massive industrial output to sustain.

While small, by 1916 the U.S. Army already had begun taking steps to turn its relatively understrength fighting force into a more efficiently run organization. Reforms—many of them pushed through by Secretary of War Elihu Root—were designed to catch up to developments in Britain, France, and Germany. They focused on promoting professionalism, building a more well-defined administrative system, and increasing training for soldiers. The most important of these reforms were the creation in 1900 of the Army War College and in 1903 of the general staff system, which was designed to manage the Army's administrative, logistical, and operational needs. Despite this change, positions and responsibilities were still often based more on negotiation and personal relationships than by well-defined, permanent administrative structures. In 1903, Congress also passed the Dick Act that reorganized the state militias into the National Guard as a reserve of the Army and mandated that its training, organization, and equipment be in line with that of the regular force. Notwithstanding these reforms, the Army remained in flux and had little experience as an organization with how to manage, command, and supply forces comprised of more than one brigade.

Elihu Root served as U.S. secretary of war, 1899–1904. (Library of Congress)

The initial 1916 expedition into Mexico consisted of 4,800 soldiers from the Regular Army. Upon their arrival at Columbus, Pershing reorganized these forces into a provisional division with three brigades—two cavalry and one infantry—plus supporting units. He divided this force into two provisional columns, an eastern column and a western column. The 2,200 officers and men at Camp Furlong formed Col. James R. Lockett's Eastern Column, which included the 1st Provisional Cavalry Brigade, consisting of the 11th and 13th Cavalry, and the 1st Provisional Infantry Brigade, with the 6th and 16th Infantry. Supporting units included Battery C of the 6th Field Artillery, Companies E and H of the 2d Battalion of Engineers, Ambulance Company No. 7, Field Hospital No. 7, Quartermaster Wagon Company Nos. 1 and 2, and some of the Regular Army's Signal Corps. Pershing initially traveled with the Western Column, which started from Culberson's Ranch, about sixty miles west of Columbus.

Apache members of the U.S. Scouts assigned to the Punitive Expedition
(U.S. Army photo)

Commanded by Col. George A. Dodd, the column consisted of the 2d Provisional Cavalry Brigade, which included the 7th Cavalry, the 10th Cavalry—comprised of "Buffalo Soldiers," African American enlisted men, commanded by mostly white officers. Battery B of the 6th Field Artillery rounded out the column—a total of 1,517 officers and enlisted men.

The Army also called on detachments of U.S. Scouts to accompany each column. Composed of Apache men recruited into the Army for their expertise in tracking and operating in the inhospitable terrain of the border region, the scouts had seen extensive service during the Indian campaigns of the previous decades. Although the number of scouts had decreased following the surrender of Geronimo (Goyaałé) in 1891, twenty-four remained on active duty in 1916. The Army attached scouts from Forts Huachuca and Apache to the 10th and 11th Cavalry for service with the Western and Eastern Columns, respectively. Although the number of scouts increased to thirty-nine during the campaign, they would see their last major operation.

To support the expedition, Pershing's quartermaster general, Maj. John F. Madden, established a support base at Columbus that soon dwarfed Camp Furlong. At the expedition's peak, the Army had built 67,025 square feet of storage, 10,266 square feet of offices, a hospital that measured 7,512 square feet, a veterinary hospital, three bridges, tanks for holding gasoline, corrals for horses and mules, and a forge. During the course of the campaign, the depot moved more than 5.3 million pounds of subsistence stores, 250,000 pounds of clothing, 138,000 pounds of mail, and other materials for the expedition's signal, engineer, and medical units. Despite the rapid increase in capacity, keeping the supply lines open to Pershing's Chihuahua headquarters at Colonia Dublán, about eighty miles south of Columbus, proved a daunting task. This problem only compounded as Pershing's soldiers moved south, requiring the quartermaster to move goods to an advanced depot at Namiquipa, about 220 miles south of Columbus. A shortage of quartermaster soldiers for such a substantial undertaking forced the Army to hire a large number of civilian laborers to unload, store, and pack materials for shipment to the forward depots, as well as a police force of forty-seven to safeguard property from theft and to maintain order among the workers.

While Pershing prepared to enter Chihuahua, Wilson turned to diplomatic channels to make the case for sending U.S. troops into Mexico in the hopes of preventing a break with Carranza's government. In this effort, Woodrow Wilson and Secretary of State Robert Lansing justified the Punitive Expedition under the diplomatic framework of "hot pursuit." Enacted during the Apache Wars of the 1880s, this concept was a response to the Apache tactic of raiding on one side of the border and then escaping pursuit simply by crossing the line into the other country's jurisdiction. The United States and Mexico signed an agreement in July 1882 to allow federal troops from either nation to cross into the other if in "hot pursuit" of an Indian band. Neither country used this agreement after 1896, but it remained within the living memory of both nations. On 10 March, Carranza's government accepted the concept of reciprocal crossings but only if "the raid effected at Columbus should unfortunately be repeated at any other point on the border." Wilson and Lansing ignored this authorization of "hot pursuit" only in the event of future raids and responded that they understood that Mexico would tolerate U.S. troops on its soil in response to the Columbus raid. Pershing's forces carried copies of a telegram guaranteeing safe passage

The Army camp at Columbus, New Mexico, which served as the Punitive Expedition's logistical support base (Library of Congress)

from Mexico's General Álvaro Obregón in order to justify their presence to Mexican troops and civilians, but the issue of the legality of the expedition remained far from settled from the standpoint of the Mexican government.

Carranza's refusal to accept the presence of U.S. troops in Mexico made it far more difficult to operate in Chihuahua and to move men and materials across the desert than it might have been otherwise. Pershing's men had to contend not just with the desert but with government troops that could be hostile to soldiers who they saw as violating Mexican national sovereignty. Carranza's reluctance to authorize the expedition also led him to deny Pershing's troops the use of Mexico's railroads after 18 March. This meant that Pershing had to rely on pack animals, wagons, and later trucks to move men and materials. The limitations on transportation led to a persistent lack of supplies during the first several months of the expedition. Major Tompkins described operating with his squadron for more than a month in the countryside with only 3 days' worth of rations and 120 rounds of ammunition. Soldiers obtained provisions beyond these limited supplies either by living off the land or purchasing goods from locals. Sometimes soldiers could pay in hard currency, often out of their own pocket, which was much appreciated by those with food to sell. But the men often were forced to pay in promissory notes that the recipient could only redeem them through unit adjutants. This almost ensured that the unfortunate Chihuahuans would never see compensation. Supply problems also meant that soldiers often went into the field with inadequate clothing. Thinking that the men would fight in the desert and that the expedition would end quickly, the quartermaster made few provisions for heavy winter uniforms, coats, or extra blankets. Many men simply had to go without these items for the first several months of the campaign.

Two days before he planned to march into Chihuahua, Pershing received intelligence that the commanding officer of the small detachment of Mexican troops charged with guarding the border gate at Las Palomas would oppose any move by U.S. forces to pass into Mexico. In response, Pershing sent a message to the border guard stating his intention to cross whether or not he met resistance and arguing that his expedition was legal. This caused some consternation in Washington because it meant that Pershing risked sparking an incident with Carranza's forces before the hunt for Villa had even begun. Despite the danger, Pershing did not receive orders to alter his march or to bypass Las Palomas. This marked the first of many occasions in which Pershing's and Carranza's forces clashed or threatened each other with violence during the course of the expedition.

Operations

The Eastern Column crossed the border on 15 March without incident. Passing through the Las Palomas border gate, it found the guard station deserted and learned that the Mexican detachment stationed there had fled earlier that morning. Problems with Pershing's touring car delayed the Western Column's departure until the following day. It reached Colonia Dublán at 1930 on 17 March, and Pershing established his headquarters there. The general chose the location for its access to the Casas Grandes River and Mexico's Northwestern Railroad, as well as its proximity to a Mormon settlement. The Mormons welcomed and readily aided the expedition by providing a place to camp, access to supplies, and sources of intelligence. In return, the U.S. Army provided protection from local bandits who preyed on foreigners living and working in Chihuahua as well as anti-American revolutionaries operating in the area. Pershing met with his staff to plan the first operations.

Acting on information that Villa had been seen about sixty miles to the south in the vicinity of San Miguel de Babícora, Pershing decided to act without waiting for the Eastern Column to arrive.

Brig. Gen. John J. "Black Jack" Pershing crossing the Río Grande into Mexico with the Western Column at the beginning of the Punitive Expedition (U.S. Army Military History Institute)

Keeping the 3d Squadron of the 10th Cavalry at Colonia Dublán, he formed three forces from the remaining units. One column, the entire 7th Cavalry commanded by Col. James B. Erwin, would advance mounted south through Galeana and El Valle and then move east in the direction of the Sierra Madre Oriental mountains in the hopes of catching Villa or his bands if they tried to ride east from Babícora. The other two detachments were to proceed to Babícora itself. Pershing decided to send these two detachments by rail at least part of the way in order to afford the troopers and their mounts a much-needed rest after deploying from Fort Huachuca, Arizona, to Colonia Dublán. The 10th Cavalry's 2d Squadron, reinforced with the regimental machine gun troop and commanded by Lt. Col. William C. Brown, would leave the train at El Rucio, a railroad station forty miles north of Babícora, and advance south. Meanwhile, the 1st Squadron, commanded by

Maj. Elwood W. Evans, was supposed to unload at Madera, about twenty-eight miles south of Babícora, and advance north. Pershing's plan sought to locate Villa with one column while the other two converged to trap him in a pincer movement. This stood as the only time during the Punitive Expedition that Pershing used the Mexican rails before Carranza's decree took effect.

Although Pershing's decision to use the railroad seemed sound, the plan proved to be disastrous. The railroad's rolling stock was in such bad condition that when the train arrived at Colonia Dublán, the soldiers had to repair doors, cut ventilation holes into the walls, and patch large gaps in the floors caused by old cooking fires to make the cars suitable for transporting men and horses. After a delay of several hours, the train departed, but fearing Villista or Carrancista interference, the conductor refused to stop in Casas Grandes, just outside of Colonia Dublán, to load firewood for the locomotive. This forced the troops to provide the necessary fuel instead by using their camp hatchets to demolish a loading pen and to chop down telegraph poles, which the U.S. government had to pay for later. The train continued only a short distance farther before the locomotive needed water and had to return again to the starting point. Once replenished, the conductor refused to proceed to El Rucio. Colonel Brown bribed the conductor to settle the matter. Finally, outside of El Rucio, the conductor found that the locomotive lacked the necessary power to pull the twenty-five loaded cattle and boxcars up the steep grade heading into town, and the train gave out short of its first destination. About 1200 on 20 March, a frustrated Colonel Brown ordered his men off the train and continued the mission on horseback. Major Evans' detachment remained on board the lightened train until two cars derailed and plunged down an embankment near Cumbre Pass, injuring eleven men. Evans then ordered the 1st Squadron to disembark and conducted a night march to Babícora, but shortly after he arrived, he received orders to join Brown. The two met outside Babícora at 1400 on 24 March. Erwin's contingent continued to patrol east of Babícora and remained out of contact with headquarters for several days.

After their rendezvous, the two squadrons learned that Villa had, in fact, not been in Babícora for several months. Suspecting he was still somewhere in the region, the squadrons continued their search mission for the next thirty-two days. Although the troopers relied on Mexican citizens they met along the trails for information, they received little useful intelligence. Peasants and

Carrancista soldiers board a train while U.S. troops wait on a railroad embankment to begin an operation against Villista forces in Mexico. (Library of Congress)

townspeople alike often viewed U.S. troops as invaders, whether or not they were sympathetic to Villa. Covering the vast Chihuahua wilderness of mountains and plains proved almost impossible given that Villistas knew the terrain and had a much greater knowledge of possible escape routes and hideouts. The soldiers also understood that while they were restricted to operating in Chihuahua, Villa and his men could simply go south beyond their reach to the state of Durango if threatened. Hearing Pershing's discussion on the location and operations of these columns at a 23 March press conference, one scout quipped, "We've got Villa entirely surrounded . . . on one side."

As the Babícora operation continued, the Eastern Column arrived from Columbus to Colonia Dublán. Just as he had with its sister force, Pershing divided the cavalry of the Eastern Column, this time into four operational detachments. Maj. Elmer Lindsley commanded the first, consisting of Troop L and the entire 2d Squadron of the 13th Cavalry. Major Tompkins led the second detachment, with Troops K and M of the 13th, and Troops I and K of the 10th Cavalry. The third and fourth detachments, under the command of Maj. Robert L. Howze and Lt. Col. Henry T. Allen, respectively, came from the 11th Cavalry.

Pershing expected these mobile units, which he called "flying columns," to operate semi-independently with minimum guidance as they gathered intelligence to locate and destroy Villista bands. Detachment commanders were to stay in communication with headquarters, but Pershing allowed them freedom of maneuver in their assigned areas without having to wait for specific orders. While Army quartermasters worked to forward provisions and other supplies, the flying columns mostly lived off the land. Major Tompkins recounted that his men frequently subsisted on a diet of beef taken from cattle they killed, butchered on the trail, and then fried in their mess kits over the campfire. The expedition's hasty assembly caused other problems besides a lack of food. Tompkins' unit, for example, had no maps, while some units used ones that dated from the Mexican War of the 1840s (*Map 3*).

Meanwhile, Pershing merged two of the three detachments from the Eastern Column, which now operated in the Babícora area. Sixty-three-year-old Colonel Dodd assumed command of his old regiment, the 7th Cavalry, as well as the 10th Cavalry. Soon after this reorganization, the 7th Cavalry encountered a small group of Carrancistas on the outskirts of El Valle, about twenty-five miles east of Babícora and fifty miles south of Colonia Dublán. Dodd requested intelligence from the Carrancista commander, Col. Enrique Salas. The Mexican officer told him that recent reports put Villa near Namiquipa, about forty miles farther south. Dodd assumed that Salas was credible and led his men in that direction only to discover that Villa had already departed. When he arrived, Dodd questioned local federal soldiers, but several people with whom he spoke gave contradictory information. Dodd next decided to lead Erwin's men south again through Santa Ana to Bachíniva, where he received orders to wait for Howze. But Dodd believed he was closing in on Villa and instead chose to press on southwest to Guerrero, about 190 miles south of Colonia Dublán.

Unknown to Dodd, Villa and his forces were in Guerrero at the time and were on the cusp of suffering a major reversal. Soon after crossing back into Mexico, Villa had ordered his men to move to Boca Grande and then to Ascensión and Galeana, about 150 miles south of Columbus. There, he attempted to reinvigorate support for his cause in the countryside and enlist recruits by giving speeches in which he promised to only attack U.S., not Mexican, citizens going forward. In one of these speeches in El Valle, he predicted that U.S. troops would soon invade Mexico with Carranza's knowledge and approval, and he warned of the damage to Mexican sovereignty.

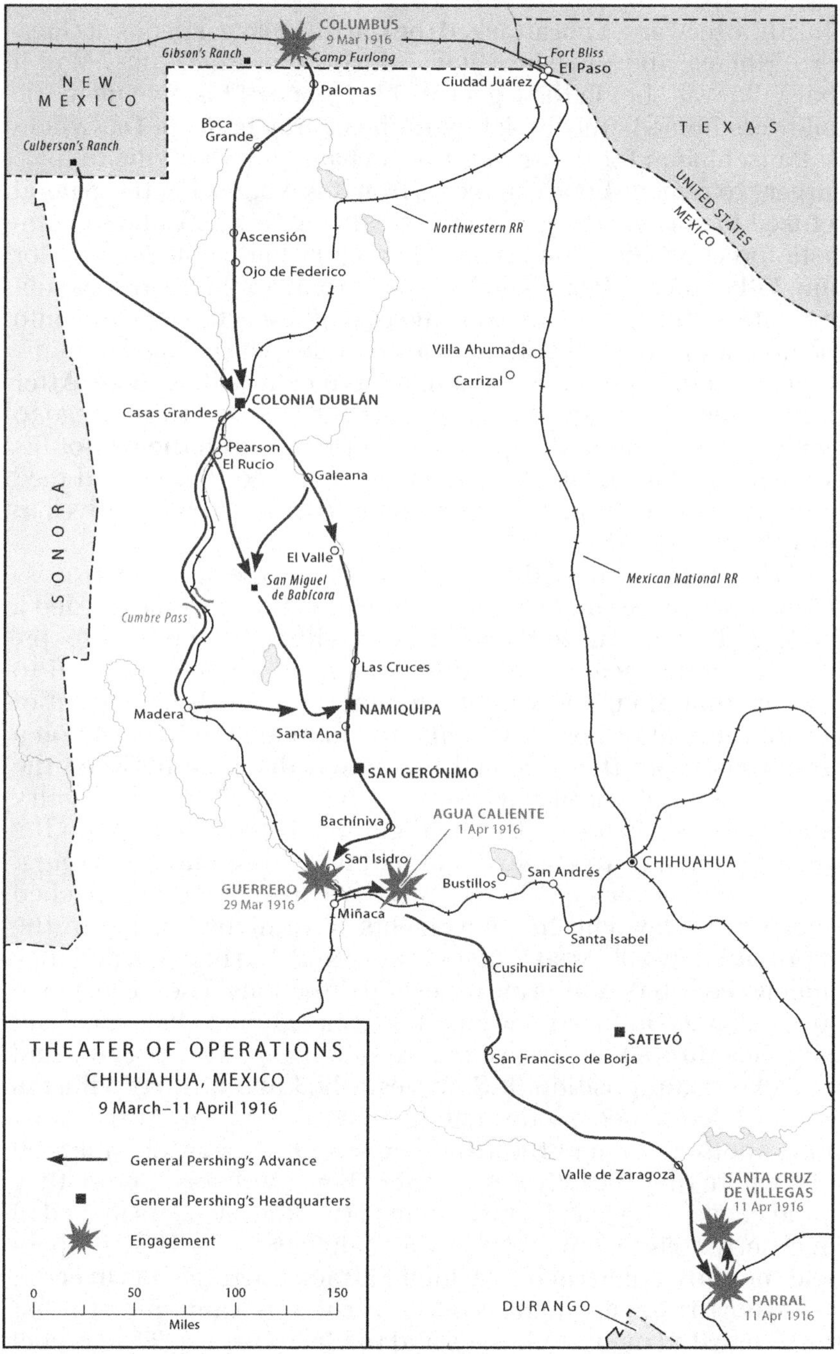
COLUMBUS
9 Mar 1916
Camp Furlong
Gibson's Ranch
Fort Bliss
El Paso
Ciudad Juárez
NEW MEXICO
Palomas
Boca Grande
TEXAS
Culberson's Ranch
UNITED STATES
MEXICO
Northwestern RR
Ascensión
Ojo de Federico
Villa Ahumada
Carrizal
COLONIA DUBLÁN
Casas Grandes
Pearson
El Rucio
Galeana
SONORA
El Valle
San Miguel de Babícora
Mexican National RR
Cumbre Pass
Las Cruces
NAMIQUIPA
Madera
Santa Ana
SAN GERÓNIMO
Bachíniva
AGUA CALIENTE
1 Apr 1916
San Isidro
CHIHUAHUA
GUERRERO
29 Mar 1916
Bustillos
San Andrés
Miñaca
Santa Isabel
Cusihuiriachic
THEATER OF OPERATIONS
CHIHUAHUA, MEXICO
9 March–11 April 1916
SATEVÓ
San Francisco de Borja
General Pershing's Advance
General Pershing's Headquarters
Engagement
Valle de Zaragoza
SANTA CRUZ DE VILLEGAS
11 Apr 1916
0
50
100
150
Miles
DURANGO
PARRAL
11 Apr 1916

Map 3

Shortly afterward, Villa attacked the Carrancista garrisons at Guerrero, Miñaca, and San Isidro. It was during this attack on 27 March that Villa was shot in the leg, probably by one of his own men. His followers immediately took him to the nearby home of Dr. Lyman B. Rauschbaum for treatment. Rauschbaum advised Villa to have surgery to remove bone fragments from his wound, but the general refused to stay for the procedure. Realizing that U.S. and Carrancista forces would soon learn of his injury and come looking for him, Villa ordered Brig. Gen. Nicolás Fernández, his principal subordinate on the Columbus raid, and a small escort to take him into the mountains to hide. Although a teetotaler who seemed immune to physical discomfort, Villa numbed the pain with alcohol. After taking shelter in a cave, Villa sent Fernández ahead to Durango to await further orders. Completely out of touch with the rest of his command—during which the rumor circulated that he had died of complications from an amputation—Villa relied on relatives to bring him food and water.

While this was unfolding, Dodd had made the decision to cross the mountains to Guerrero, but to do this efficiently, he needed a guide. At first, no one in Bachíniva was willing to help U.S. troops, but after much searching, Dodd convinced a reluctant local to lead his unit along the winding mountain trails. Whether out of fear of reprisal or loyalty to Villa, the man led the troopers on a circuitous route that doubled the actual distance between the two towns to about fifty-five miles. Then, when the 7th Cavalry finally neared Guerrero, the local guide refused to advise on the best approach to town, so Dodd had no choice but to surround it and wait for daylight to advance. Because Dodd approached Guerrero in daylight, his movements were highly visible to the approximately 200 or so Villistas who remained in town, and they quickly retreated, scattering in several directions. The cavalrymen pursued, and Dodd led one small contingent on a chase of about ten miles through an arroyo before U.S. troops set up a rifle and a machine gun position. U.S. forces killed about thirty Villistas as they tried to escape through the narrow canyon. At the same time, another troop of the 7th Cavalry spotted a group of armed men whom they initially suspected of being Villistas. Before they could engage, the cavalrymen noticed that the suspects moved in an orderly column with a large Mexican flag, so the U.S. troopers held their fire concerned they might attack Carrancistas in error.

After the engagement, Dodd returned to Namiquipa to rest the exhausted men and horses and to file his report. The colonel noted that because many of the Villistas in Guerrero were Yaqui Indians, making an accurate count of enemy casualties was difficult. The Yaqui tended to carry their dead off the battlefield in order to obscure their true losses. He concluded that his men therefore probably killed or wounded more of the enemy than the corpses left behind indicated. Dodd did not know, but at the time of the skirmish, Villa rested only a few hours' ride outside of town and was moving slowly due to his injuries. Dodd's attack and pursuit at Guerrero proved the closest anyone in the Punitive Expedition came to capturing Villa.

Col. George A. Dodd, pictured after his promotion to brigadier general (National Archives)

The clash at Guerrero further convinced Dodd that questioning Mexican soldiers and citizens was unlikely to yield useful intelligence. Not only were local people unwilling to help in many cases, but Villa often kept his location a carefully guarded secret, even from his own officers. Often, neither locals nor Villista soldiers knew where Villa was operating. Carrancista forces experienced similar problems trying to get reliable intelligence, but even when they came across credible information, they were reluctant to share it for fear that U.S. operations would hamper their own. Pershing likened it to "interfering in a family quarrel; they would fight each other when we were not about, but when we appeared frustration of our plans was the common objective."

Pershing's forces had to develop alternatives to gathering intelligence on the ground from Mexican citizens. Maj. James A. Ryan, Pershing's intelligence officer, led this effort, and during the first several months of the expedition, he assembled a network of trusted local informants and civilian scouts. Their work helped to compensate for the shortage of reliable maps and supplemented the tracking efforts of U.S. Scouts, not only in locating Villa's men, but in finding animals, fodder, and water to add to their meager supplies. Ryan also transmitted numerous reports by telegram to the War College Division of the General Staff and its Military Information Section, although no evidence exists that anyone acted on or even read them.

Parral

Hearing that Villa had gone into hiding due to his injuries several days after the clash at Guerrero presented Pershing with a serious dilemma. It now seemed like it would be even more difficult to find information on Villa's whereabouts because the general left few clues indicating where he might be holed up, and when asked, the answers locals provided "would have indicated Villa's departure in almost any direction and his presence in several places at once." Knowing his men would have to push farther south to continue the search, Pershing moved his headquarters from Colonia Dublán to San Gerónimo, 140 miles to the south. He arrived there on 27 March during a sandstorm that coated the camp and all the equipment in a layer of fine sand and grounded all of the 1st Aero Squadron's planes. In the absence of any concrete intelligence, Major Tompkins suggested to Pershing that Villa could be in the vicinity of Parral, about 150 miles south of Guerrero, near the border with Durango, because he had used that town as a base before the revolution. Pershing concurred and planned an advance using parallel columns from Guerrero to Parral. Colonel Brown led his column along the route farthest east with Tompkins in the center and Major Howze on the westernmost trail. When Tompkins set out, he took twelve pack mules to carry supplies, 500 pesos worth of Mexican silver, and rations for five days. Before leaving San Gerónimo, the columns supplemented their austere supplies through local purchase and foraging. Meanwhile, Pershing sent Colonel Allen from Colonia Dublán to Namiquipa. His mission was to ride another 175 miles south and east to Satevó to look for Villa's subordinate, General Pablo López.

The advancing detachments encountered both Carrancistas and Villistas. On 1 April, Brown's men unexpectedly ran into a group of 150 Villistas commanded by Col. Francisco Beltrán in the mountains near Agua Caliente and forced them to retreat after a brief skirmish. Maj. Charles Young of the 2d Squadron,

10th Cavalry, one of the few African American officers in the Army, led a charge against one of the groups of retreating Villistas, then pursued them for seven miles in an action that killed two enemy with no U.S. losses. The day after their skirmish in Agua Caliente, Brown's men encountered Carrancista General José Cavazos, who warned them to proceed no farther into Chihuahuan territory. Three days later, Howze's column also encountered Cavazos as it traveled south over the mountains through the Gabilana trail. The approximately 200 Carrancistas under Cavazos' command drew their rifles in preparation to charge when Howze rode between the lines and identified his men as U.S. Army. Cavazos' men halted, and the U.S. troopers continued south. Meanwhile at headquarters, General Luis Herrera visited Pershing. The Carrancista officer told the U.S. general that he felt certain Villa was dead, making the expedition's continued presence unnecessary. When Pershing indicated that he would continue pursuing Villa and his bands, Herrera warned him not to continue farther south lest it result in a confrontation between Carrancistas and U.S. cavalry. Pershing believed that Herrera came to his headquarters at San Gerónimo less to caution him and more to gather intelligence about U.S. forces. He therefore ignored the Carrancista's warning and moved his headquarters even farther south to Satevó.

Maj. Charles Young commanded 2d Squadron, 10th Cavalry, on the Punitive Expedition. Pictured shortly before promotion to lieutenant colonel, Young was the third African American to graduate from West Point. (National Archives)

Moving toward Parral on 5 April, Tompkins' 100 men also encountered Cavazos and his troops. The general told Tompkins he believed Villa was dead and buried in the vicinity of Santa Ana, and he warned Tompkins to stop advancing south.

Cavazos added that the danger was not only that Carrancistas would clash with U.S. troops but that he would not be able to prevent hostile civilians from attacking U.S. soldiers. Believing Villa was still alive, Tompkins simply bypassed Cavazos' position and continued south despite the warning. Riding through Rancho Cienegitara, the major learned that Villistas had robbed the town of 2,500 pesos the day before, confirming to him that he remained on their trail. Tompkins' troopers further validated his suspicions when they surprised a group of Villistas in the act of robbing a small factory in Valle de Zaragoza on 10 April. They intervened, gave chase, and then returned the goods the Villistas had discarded in flight.

As Tompkins' men made camp the night of 10 April, Capt. Antonio Mesa approached and identified himself as a member of the federal army garrison at Parral. He offered to notify his commander that Tompkins' troops were coming, so that he could escort the U.S. cavalrymen into town. Mesa also promised that the cavalry could camp, graze their horses, and replenish their supplies at Parral. Tompkins accepted, but when the cavalry arrived on the outskirts of town, they saw no welcome party from the garrison. Wary, Tompkins approached a guard post and asked for an escort to go see the garrison commander. These men took him to General Ismael Lozano, who expressed surprise that the 10th Cavalry was in Parral and denied hearing anything about their arrival from Mesa. Lozano then told Tompkins he should not have come so far south and warned that the local population, hostile to the expedition, might present a danger to his men's safety. The major dismissed these concerns but declared he would leave the next day if Lozano designated a place for his men to camp for the night. He also mentioned that he needed supplies, and Lozano sent for an English-speaking merchant to provide provisions and fodder. As they spoke, Tompkins watched through a window to the street below as someone turned loose a mule to charge into his troops. By the time he had finished with Lozano and prepared to leave, a crowd had gathered in the plaza as a woman whipped them into a frenzy shouting, "Viva Villa!" and "Viva México!" Led by a man whom Tompkins at the time suspected was German without any particular foundation, the crowd followed the U.S. cavalrymen out of town. Though no evidence exists that foreign agents incited the mob, the presence of this man caused the major to suspect Germany of a covert plan to cause a wider diplomatic rift between the United States and Mexico.

Lozano next led Tompkins and his detachment to a campsite east of town, but Tompkins rejected it. Hemmed in by a horseshoe-shaped ridge, he deemed it impossible to defend. As a precaution, Tompkins sent a group to occupy the hill to their west. He then realized that someone in the crowd that had followed them from town was firing on his rear. At the same time, a force of about 550 Mexican soldiers began assembling on the hills across the road about 600 yards away. Tompkins demanded an explanation but saw Lozano's pistol still in its holster, and he decided that the general was not to blame. The Carrancista officer tried to halt the advancing crowd but demanded that Tompkins leave immediately. Tompkins again refused but saw the Carrancistas moving against his left flank and sent a patrol to intercept them while yelling at the Mexican mob to stay back. Someone fired another shot in the major's direction, killing Sgt. Jay Richley, and an exchange of gunfire followed. After two more men were wounded, Tompkins ordered the men to withdraw.

Once his men reached the Parral–Santa Cruz de Villegas road, Tompkins decided to ride north to Santa Cruz, about three hours away. The Mexican soldiers followed and fired on the withdrawing U.S. cavalry, while Lt. Clarence Lininger and an eight-man rear guard returned fire. As the enemy drew closer, Tompkins ordered Lininger and his men to dismount and take positions behind a stone wall to slow the enemy advance. Staying with them to observe and hold the reins of his men's horses, Tompkins suffered a gunshot wound in the shoulder. Meanwhile, the lieutenant braved enemy fire to ride back and rescue a dismounted trooper before the enemy could capture him, an act for which he later received the Distinguished Service Medal. When his men crested a ridge as they withdrew, Tompkins saw an opportunity to put some distance between them and the Carrancistas. He ordered twenty troopers to dismount and form a skirmish line on the reverse slope. They opened fire and killed forty-five Mexican soldiers as they crossed over the top and came into view. The stunned Carrancistas ended their pursuit, and the cavalrymen continued without incident to Santa Cruz, having suffered two dead and six wounded. In his report, Tompkins attributed the light losses to poor enemy marksmanship due to their paucity of serviceable firearms, inadequate training, and shortages of ammunition resulting from years of war and revolution.

From Santa Cruz, Major Tompkins sent messengers north to find and bring reinforcements from the 10th Cavalry. They encountered Capt. George Rodney, who at first suspected the messengers were deserters, but after they explained the situation, Colonel Brown ordered Major Young and the 2d Squadron to Tompkins' relief. Lozano arrived in Santa Cruz later that day and sent Tompkins an official note that stated, "I beg you to retreat immediately, in order to avoid hostilities. Otherwise, I will be obliged to attack you with the forces in my command." Major Tompkins stalled for time and replied with a written protest over the Parral skirmish. He agreed to withdraw from Santa Cruz only if guaranteed safe passage. Colonel Brown and Major Young arrived with the reinforcements about 2000 on 12 April and bivouacked outside of town. Accompanied by General Luis Herrera, General Lozano rode out to see Brown and warned that he should return north because his continued presence just strengthened popular support for Villa in the region. When Brown refused to make any commitments, Herrera became angry and stated that U.S. forces would only be allowed to move north, not east, west, or south. Brown defiantly responded, "That flag does not move one step north until I have orders from my commanding general to move it."

Pershing heard about the skirmish two days after the clash when Capt. Benjamin D. Foulois of the 1st Aero Squadron delivered a communiqué from the U.S. consulate in Chihuahua. Lozano had sent a notice to the consulate soon after the skirmish, but it was translated poorly, leading to temporary confusion in Pershing's headquarters. After receiving clarifications from Dodd, Pershing decided to withdraw Dodd, Brown, and Young's forces from the Santa Cruz area north and to move his headquarters north to Namiquipa in order to better communicate with Washington via the wireless relay at Casas Grandes. In the days after this withdrawal, Pershing sent General Luis Gutiérrez a written protest concerning what happened at Parral and submitted his revised plan of operation to General Funston. Pershing argued that his troops found "obstruction, changing more and more to open hostility the farther south we progressed," and he recommended "the immediate capture of the city and state of Chihuahua and the seizure of all railroads therein."

Pershing had some support in the Army for escalating the conflict. Maj. Gen. Tasker H. Bliss, assistant chief of staff of the Army, thought that Pershing's limited intervention into Mexico was unlikely to produce results, and he urged President Wilson to prepare a full-scale invasion to pacify the country. General Scott recommended that the president at least call the National Guard into federal service and position several thousand men on the border to deter future raids. The president rejected both recommendations. He had become anxious about committing more forces to capture Villa when involvement in the war in Europe seemed more likely. The most recent plans that Wilson could reference—written in 1911, by then-president of the Army War College Brig. Gen. William W. Wotherspoon—for a hypothetical invasion of Mexico estimated that an occupation of the whole country would require a force of up to 200,000 troops. Wilson knew that the United States could not do this and still send troops to Europe at the same time, so instead the president chose to use the Parral crisis to call for negotiation with Mexico. Secretary Lansing sent Carranza a proposal for negotiations on 22 April, and Generals Scott and Álvaro Obregón began talks in Ciudad Juárez six days later. Although neither side wanted the conflict to escalate, the agreement that the generals outlined collapsed when Obregón refused to accept a promise of U.S. withdrawal without a definite date specified. Negotiations ended amicably, however, leaving hope that talks would resume in the near future.

After reestablishing his headquarters at Namiquipa, Pershing again reorganized his forces. The 5th and 17th Cavalry, and the 24th Infantry, had recently joined the expedition, and Dodd's and Brown's detachments were moving north and needed to be assigned new areas to patrol. The general decided to divide northern Chihuahua into five districts, each centered on a principal town garrisoned with a regiment of cavalry, reinforced with artillery and infantry. Pershing noted that the Villistas no longer operated as a large, cohesive body led personally by their general but were organized into small, scattered, quickly moving bands commanded by Villa's subordinates. The organization of districts allowed the cavalry to conduct sustained patrols that ranged in size from units of 20 to 300 men who would become more familiar with the terrain, cultivate intelligence sources, and create relationships with local government and military leaders. The five districts were Namiquipa, 10th Cavalry; Guerrero, 7th Cavalry; Bustillos, 13th Cavalry; Satevó, 5th Cavalry; and San Francisco de Borja, 11th Cavalry.

Cusihuiriachic

As Pershing's operations continued, Villa's forces launched a series of small-scale attacks. On 4 May, Generals Julio Acosta and Cruz Domínguez led an assault on the Carrancista garrison at the largely Mennonite silver mining town of Cusihuiriachic, near present-day Ciudad Cuauhtémoc. The next day, the local Carrancista commander requested assistance from the U.S. cavalry to aid in dealing with the emergency. Major Howze led a reinforced squadron of six lettered troops, a machine gun platoon, and a detachment of Apache scouts from the 11th Cavalry, a total of 333 officers and men, on a night march to Cusihuiriachic. The 140 Villista raiders left before Howze arrived, but the detachment pursued the enemy into the mountains with the assistance of local guides. As the retreating Villistas approached Ojos Azules ranch, they stopped and opened fire on Howze's men. Troop A then drew pistols and charged, while the rest of Howze's force circled the objective and took blocking positions to prevent the enemy's escape. The U.S. cavalry freed four Carrancista prisoners and killed forty-two Villistas in the fight, but General Acosta, General Domínguez, and most of their men scattered into the desert. Howze's detachment lost a few horses but suffered no casualties in the lopsided victory.

Glenn Springs and Boquillas

One day after the Cusihuiriachic raid, a group of about a hundred Villistas led by Lt. Col. Natividad Álvarez crossed the border into the Big Bend region of Texas. Álvarez divided his force, personally leading one division toward the town of Boquillas, Mexico, and sending the other under the command of one of his lieutenants to attack Glenn Springs, Texas. In Boquillas, Álvarez and his men met no resistance. After looting several buildings, the soldiers robbed and captured Jesse Deemer, a U.S. citizen living in Mexico and manager of the local general store, and his Afro-Seminole clerk, Monroe, or Maurice, Payne. Across the border, Glenn Springs sat as an isolated village of about eighty inhabitants, mostly Mexican Americans, fifty of whom worked in a local wax extraction factory. The town was also home to Sgt. Charles E. Smyth and his nine-man squad from Troop A of the 14th Cavalry. As the attack began, they rushed from their tents and took cover in an adobe building before engaging in a night-long gun battle. They only vacated their position when Villistas set the building on fire. The troopers then continued to fight from a nearby hill and suffered three dead and four wounded. Sergeant Smyth later received the Distinguished Service Medal for this action. During the clash, the Villistas ransacked the wax factory, looted the general store and several homes, and set part of the town ablaze. About 0300, they gathered their wounded and dead, except for two, and rode

The Machine Gun Troop, 8th Cavalry, conducting a charge during a training exercise near its post in Texas near the Mexican border (Library of Congress)

off to Boquillas to meet the other division. The two detachments met up outside of Boquillas at 1000 and moved south, but in the confusion of the withdrawal, some miners managed to capture Álvarez. Moving further into Mexico, one small group of soldiers halted to rob the payroll and company store of the U.S.-owned Del Carmen Silver Mine. There they took four U.S. citizens prisoner, but the captives escaped after they purposely crashed their vehicle into a ditch. Scott ordered Funston to pursue the Glenn Springs raiders and to rescue the remaining two hostages.

On 7 May, Col. Frederick W. Sibley of the 14th Cavalry left Fort Clark, Texas, along with his regiment's headquarters, machine gun troops, and Troops H and K. He arrived at Glenn Springs on 11 May and was joined by Maj. George T. Langhorne, with Troops A and B of the 8th Cavalry from Fort Bliss. The major led the two 8th Cavalry troops, with two wagons and a touring car, across the Río Grande at 2000 that night, while Sibley followed with the 14th Cavalry units at 0600 two days later. After a 24-hour march, one of Langhorne's troops encountered a group of Villistas near the village of El Pino, where Deemer and Payne, the two remaining captives, were being held. The U.S. troopers charged, and the Villistas fled without firing a shot. The two hostages were recovered.

Soldiers of the 24th Infantry guarding Villista prisoners in Mexico
(Library of Congress)

Pursuing units continued searching for Villistas, and on 15 May a patrol from Troop B of the 8th Cavalry, commanded by Lt. Stuart W. Cramer Jr., killed five Villistas and wounded two without suffering any casualties in a firefight near Castillón. Cramer's men recovered nineteen horses, mules, and a wagon full of weapons, saddles, and other goods. With their mission to rescue the hostages complete and with Carranza complaining to Wilson through diplomatic cables that this second expedition could trigger a war between the two nations, General Scott ordered Sibley to return to the U.S. side of the Río Grande.

Beyond the disruption of the incursion itself and Sibley's expedition across the border, the Glenn Springs raid led Wilson to implement two major changes that each had a far-reaching effect on the Punitive Expedition. First, Wilson realized that far from being secure due to the large presence of U.S. troops, the border remained vulnerable to attack from the south. As a result, on 9 May he called the National Guard from three border states to guard the boundary between the United States and Mexico. In response to the call-up, Carranza declared his government would view any increases in the number of U.S. soldiers within the borders of Mexico as a deliberate act of war. The second action that Wilson took in the wake of Glenn Springs was to order Pershing to move his headquarters from Namiquipa back to Colonia Dublán, about 120 miles to the north. The cavalry remained responsible for patrolling its districts, but moving the headquarters to the far north of the state meant that it would be much easier going forward for both Carrancistas and Villistas to avoid U.S. forces in the field.

Pershing registered his discontent with the War Department over this withdrawal to the north and repeated his request for authorization to establish a U.S. military government for all of Chihuahua. This would have drastically altered the mission of the Punitive Expedition. Since Carranza's forces at least nominally controlled Chihuahua, a U.S. takeover of the state would have transformed Carrancistas into the expedition's primary belligerents, taking focus off of the Villistas responsible for attacking Columbus. Pershing's forces already had clashed with Carrancistas at Parral, and Mexico's federal forces increasingly were seen as secondary opponents, but President Wilson balked at declaring war against the officially recognized government of Mexico.

Carrizal

Despite the move to the north, Pershing's men continued to engage in minor clashes with Carrancistas and Villistas in Chihuahua. On 14 May, Lt. George S. Patton Jr. led a fifteen-man detail in three automobiles to purchase corn. When Patton and his men arrived at San Miguel Ranch, twenty-five miles south of Colonia Dublán, a party of horsemen rode through an arched gateway and fired on them without warning. The U.S. cavalrymen returned fire and killed three of the attackers, one of whom turned out to be one of Villa's principal subordinates, Col. Julián Cárdenas. Later in the month, Sgt. James M. Mayson led a detail of ten men from the 17th and 24th Infantry at Namiquipa on a mission to map the Santa María Valley and hunt for cattle. About six miles from the town of Las Cruces, a party of thirty Villistas rode toward a group of the surveyors and started shooting. Mayson sent one man for reinforcements while he and Cpl. Earl Philips stayed with the other eight men and fought until troopers of the 11th and 13th Cavalry arrived, and the enemy retreated. U.S. forces lost one dead and three wounded but killed Col. Candelario Cervantes, yet another chief Villa subordinate.

Recognizing the strained relations between his forces and Carrancista troops operating in the area, Pershing invited General Gabriel Gavira, the Carrancista commander responsible for the Ciudad Juárez area, to meet with him at Colonia Dublán. On 1 June, the two men talked for two hours in Gavira's railway car, and the Mexican general agreed to reduce the number of Carrancistas posted along the Northwestern Mexican Railroad from the border to Colonia Dublán. Fewer Mexican troops in the area further reduced the likelihood of clashes between the two forces.

General Treviño, the Carrancista commander of the 1st, 3d, and 5th Divisions of the Army of the Northeast, was less conciliatory. On 16 June, he relayed a terse message to Pershing's camp that read:

> I have orders from my government to prevent, by the use of arms, new invasions of my country by American forces and also to prevent the American forces that are in this state from moving to the south, east, or west of the places they now occupy. I communicate this to you for your knowledge for the reason that your force will be attacked by the Mexican forces if these instructions are not heeded.

Pershing replied, "I shall use my own judgment as to when and in what direction I shall move my forces in pursuit of bandits or in seeking information regarding bandits." Treviño already had begun his offensive against Villistas in Chihuahua, moving his headquarters to Chihuahua city on 16 May. Thus, he was motivated to restrict Pershing's movements because he did not want U.S. forces to interfere with his operations.

In late June, Pershing received reports that a large group of Carrancista soldiers had assembled near Villa Ahumada, about seventy miles east of Colonia Dublán. Pershing sent Capts. Charles T. Boyd and Lewis S. Morey, commanding Troops C and K of the 10th Cavalry, respectively, to investigate. Troop C moved from Colonia Dublán east to Ahumada, while Troop K marched from Ojo de Federico, about fifty-five miles north of Colonia Dublán, southeast to Ahumada. Boyd assumed overall command when their paths crossed at Santo Domingo ranch. There, W. P. McCabe, the U.S.-born ranch foreman, told them about recent Carrancista movements in the local area.

Pershing had warned Boyd of the delicate state of relations between the United States and Mexico before he left headquarters and advised him to avoid clashes with Carrancistas at all costs. This was to be a reconnaissance mission only. Once on the trail to Ahumada, however, Boyd confided to Lem Spilsbury, his Mexican-born

Buffalo Soldiers of the 10th Cavalry on the trail of Pancho Villa in Mexico during the Punitive Expedition (National Archives)

Mormon guide, that if Carrancista troops fired on them, Pershing was ready to move south, and General Funston would send more troops into Chihuahua. To Spilsbury, Boyd seemed to be hinting that the scouting mission had a greater purpose. Pershing later explained that Boyd must have believed that if he engaged with Mexican troops they would withdraw rather than fight. In Pershing's view, Boyd simply underestimated the Carrancista army.

With Troops C and K combined, Boyd ordered the entire force of three officers and eighty enlisted men to march to Villa Ahumada through the town of Carrizal. Spilsbury stood skeptical of the idea. McCabe warned against the plan because of the town's many bulletproof adobe buildings, which provided an advantage to the Carrancista defenders in case of a battle. Morey voiced his concerns as well, noting that while the orders read to go to Ahumada through Carrizal, caution dictated that they should bypass town, rather than force their way through the settlement. Boyd rejected these concerns and remarked about "making history." He deposited his valuables with McCabe for safekeeping and sent a courier to Carrizal, asking permission for his troopers to move through the town.

Boyd's detachment left Santo Domingo ranch at approximately 0400 on 21 June. About a mile and a quarter from the ranch, the troopers encountered a force of around 400 Carrancistas positioned on a low irrigation trench west of Carrizal. Boyd and Spilsbury rode ahead and met two Carrancista officers, General Félix U. Gómez and Maj. Genevevo Rivas. Gómez firmly stated he had orders not to allow U.S. troops to move in any direction except north. The Mexican officer said he had forwarded Boyd's message asking to enter the town to his superior, but General Treviño had denied him permission. Spilsbury again advised Boyd to march around rather than through Carrizal. Boyd disagreed, using a tone that even Gómez, who did not speak English, understood. Gómez replied, "You may go through, but . . . over our dead bodies."

Boyd led in front as the Buffalo Soldiers advanced in a "line of foragers" formation southeast toward Carrizal. First Lt. Henry R. Adair commanded Troop C on the left of the road, and Captain Morey and Troop K moved on the right. When they had advanced to within 500 yards of the Carrancistas, the troopers dismounted, sent their horses to the rear, and continued forward on foot on a line. At the same time, Mexican cavalry began moving around their flanks. When Boyd's forces came to within 200 yards, the Carrancistas opened fire with rifles and machine guns. The men of Troop K moved into a small depression on their side of the road and took up prone firing positions but lost contact with Troop C. Captain Boyd lay dead after receiving three gunshot wounds; Adair also was killed in the fighting. Morey, despite a shoulder wound, assumed command and ordered a retreat. Most of the men began streaming northwest toward Santo Domingo ranch. Morey tried to continue with them, but heavy blood loss forced him to hide in a lime kiln with four other wounded men until nightfall when they continued their slow march to the ranch. They eventually found Major Howze, who by then had been dispatched from Colonia Dublán to look for survivors. Meanwhile, McCabe, the foreman at Santo Domingo, watched the surviving troopers straggle in from the skirmish and on toward headquarters at Colonia Dublán. He and his remaining hands left the ranch the next morning before a group of fifteen or twenty Carrancistas arrived, looking to capture retreating U.S. cavalrymen (*Map 4*).

The skirmish routed Troops C and K. The survivors estimated they had wounded forty and killed forty-five Mexican troops, including General Gómez, but the U.S. cavalrymen lost nine men and suffered twelve wounded and three missing. Twenty-three troopers, as well as their civilian guide, were taken prisoner. The Carrancista forces moved their captured by train to Chihuahua

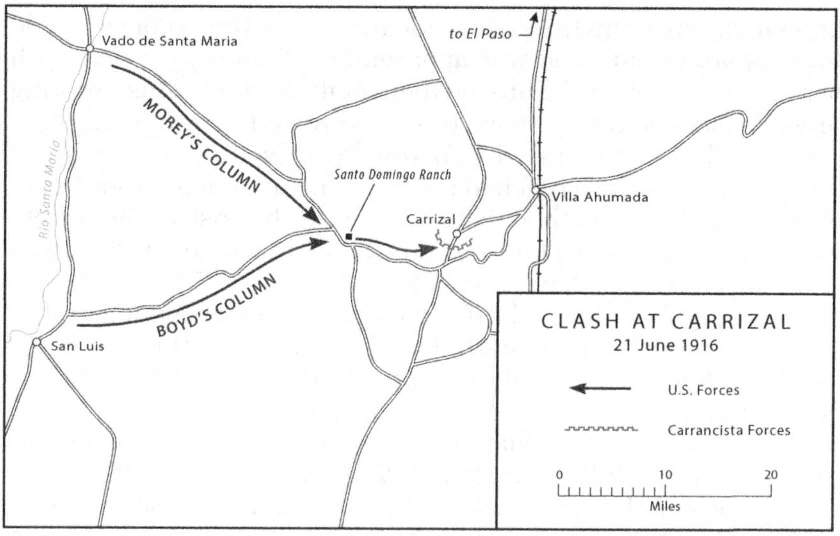

Map 4

city and imprisoned them in the city penitentiary. Although crowds briefly pelted the captives with stones during their march through town, they were treated well in prison and were visited daily by the British consul and by an Associated Press reporter. The U.S. reading public stayed informed about the prisoners' condition throughout their captivity. After several terse notes back and forth with Wilson, Carranza finally agreed to release the prisoners as a gesture of goodwill. They traveled to El Paso by train on 29 June, with all in fairly good health.

Pershing was slow to learn about the engagement at Carrizal. He only discovered that his men had fought with General Gómez's troops on 22 June when the first survivors straggled back to Colonia Dublán. By then, Funston already had heard the news. Carrancista forces at Carrizal telegraphed reports of the skirmish to Chihuahua city soon after the end of the fighting. The Mexican press reported on the engagement the same day, and the U.S. press followed. President Wilson first learned about the skirmish by reading about it in the evening newspaper on 21 June. An infuriated Funston immediately telegraphed Pershing demanding an explanation for his lack of communication with his superiors in the United States. "Why in the name of God do I hear nothing from you?" he roared. He told Pershing, "The whole country has known for ten hours through Mexican sources that a considerable force of your command was apparently defeated yesterday with heavy loss at Carrizal." Funston then reminded him of his "existing orders" and asked why those troops were so far from Pershing's position. The exasperated Funston expressed dismay that the news seemed to have reached Pershing "nearly twenty-four hours after the affair" even though he was relatively close to the action. Funston demanded to know "who was responsible for what was on its face seems to have been a terrible blunder?"

Working from limited information, Pershing responded that Mexican troops had committed an act of war by attacking Boyd and his soldiers. He repeated his recommendation that his forces conquer the entire state of Chihuahua and seize the railroads. Funston informed him that he could not authorize such a move until there was a full investigation of the incident. The subsequent testimonies of Morey, McCabe, and other survivors made it clear that Boyd, not the Carrancistas, had instigated the fight. Despite this revelation, Pershing still blamed the Mexican forces and thought their actions at Carrizal constituted an act of war.

Logistics Advances During and After the Expedition

Carrizal marked the effective end of the campaign to find Villa. In the immediate aftermath of the skirmish, Funston ordered Pershing to stop sending patrols beyond an arbitrary zone with a boundary of 150 miles south of the U.S.-Chihuahua border. In practice, this confined Pershing's forces to the area around Colonia Dublán. While Pershing's men no longer campaigned in the field, the 7,000 men at Colonia Dublán and the 4,000 at nearby El Valle still had to be fed and provided with supplies. This presented a major logistical challenge because Pershing could not use Mexico's railroads to move men or goods. As a result, the soldiers of the Punitive Expedition came to rely heavily on trucks to meet the troops' daily demands. At the same time, Pershing used airplanes to do basic reconnaissance and to maintain communications. The Punitive Expedition therefore served as a technological proving ground as the Army adapted new, untested technologies to military operations in a challenging environment. The lessons learned enabled the Army to close some of the technological gap with its European counterparts and implement improvements that translated to operational success on the battlefield over the course of the next two years.

Prior to the Punitive Expedition, the Army mostly relied on mule-drawn wagons and pack trains to meet the logistical demands of its soldiers in places the railroads could not reach. Each Army escort wagon, drawn by a team of four mules, could haul 3,000 pounds of cargo over flat terrain on good roads with no steep grades under ideal weather conditions. If mules had to draw wagons uphill or negotiate uneven roads, as they did in Chihuahua, the men driving them had to reduce the weight they carried. The wagons also had to carry fodder to feed the animals, which further reduced the amount of cargo they could transport. Wagons were clearly not an efficient means of moving goods and material in northern Mexico, but the alternative—loading cargo onto the backs of mules—also proved unable to meet the needs of Pershing's soldiers. A mule with a pack saddle carried a maximum load of 200 pounds, but due to the sparse terrain of northern Mexico much of this space had to be used for animal fodder, reducing available space for unit supplies.

Given the limitations on the use of draft and pack animals, General Pershing and his quartermaster, Major Madden, had few options but to employ trucks for transport. This presented a challenge as the Army owned few trucks, and most of these had been purchased for use at local posts. As late as March 1916, the entire Southern Department had just 16 trucks, while the Army as a whole had 56 automobiles and 105 trucks in service. Of these, fifty-four were sent to the border at the start of the expedition, providing Pershing with two truck companies. Unfortunately, these trucks arrived at Columbus with the chassis detached from their bodies for easier transport. Upon their arrival, the men of Truck Companies 1 and 2 discovered the shipment did not include the required bolts to reattach the chassis. As a result, machine-shop personnel and volunteers from the 1st Aero Squadron and the Signal Corps worked thirty-six hours nonstop to prepare the trucks for service. After this initial ordeal, support personnel became more proficient in quickly assembling the components of the next two shipments, which arrived on 28 March and 1 April 1916. Though the men could quickly reassemble the trucks, they still had to contend with repairing a wide array of trucks of different makes and models without the benefit of having standard parts or components.

With so few trucks available for the expedition, General Scott, the Army chief of staff, became concerned about supply bottlenecks forming at Columbus. He asked Army Quartermaster General Maj. Gen. James B. Aleshire if the Army had sufficient trucks

The Army that deployed to the border was a force in transition. Quartermaster units used both old and new modes of logistical support, employing mule-drawn escort wagons as well as motorized truck transport, to support the campaign on the border and in Mexico. (Library of Congress)

to supply Pershing with the necessary materiel. When Aleshire confirmed that they did not, Scott asked the quartermaster to estimate the amount of money needed to purchase additional motorized vehicles. Aleshire offhandedly replied it would cost $450,000, and much to his surprise, Scott ordered him to spend the money on trucks, mechanics, and maintenance facilities and to send them all to Columbus. The chief of staff then approached Secretary Baker and said, "You will have to use your good offices with the President, Mr. Secretary, to keep me out of jail." After Scott explained what he had just spent on trucks without an appropriation from Congress, Baker casually responded, "Ho! That's nothing! If anybody goes to jail I'll be the man—I'll go to jail for everybody." Over the eleven months of the expedition, Pershing's forces expanded from two truck companies to seventeen quartermaster truck companies carrying supplies from Columbus to the advanced bases in Mexico. The newly purchased vehicles included 588 cargo trucks, 57 tanker trucks, 12 truck-mounted machine shops, and 6 wreckers.

This increase in the number of trucks on the border represented such a major expansion that the Army initially had to hire a large number of civilian drivers and mechanics. Because driving, maintaining, and repairing the large variety of models on extremely poor dirt roads proved physically demanding, the Army had to pay $100 per month to attract the number of civilians needed to drive all of the trucks. Given that the average soldier earned $21, the disparity caused some friction. After experimenting throughout the expedition, the Quartermaster Corps determined that to achieve maximum efficiency each truck company required a company commander, one subaltern, one truckmaster, nine assistant truckmasters, nine drivers and nine assistant drivers, one machinist, one assistant machinist, one clerk, one mess steward, and two cooks.

The improved availability of trucks and their organization into convoys helped to ease but did not solve the problem of moving supplies to Colonia Dublán and beyond. Chihuahua's exceedingly poor roads made it difficult to establish supply lines as U.S. cavalrymen moved into remote areas. Pershing's decision to relocate his headquarters south during the first weeks of the expedition proved to be particularly taxing because truck companies now had to maintain a supply line of more than 400 miles into areas where the revolution left roads in extreme disrepair. At first, the expedition's engineer battalion focused on repairing roads in northern Chihuahua that had been destroyed by fighting, but as operations dragged on it became clear that new roads and bridges would have to be built. This task proved more challenging than expected. The engineers dug roads in the sand only to find them reburied after the next wind storm or swallowed by mud pits following a drenching rain. As a result, they began laying gravel roads. By the end of the campaign, Army engineers had built 157 miles of new roads and repaired 224 miles of existing ones at a cost of $380,000. Members of the Corps of Engineers also participated in mapping expeditions that improved the soldiers' knowledge of Chihuahua's topography.

Keeping the expedition supplied remained a challenge, but despite the cost of purchasing new trucks and building roads, motorized transport proved more cost-efficient than General Aleshire anticipated. With the ability to operate under adverse conditions, trucks saved money per mile when compared to wagons and pack animals, and over the long term, they promised to be even more efficient as the Army trained its own drivers and learned to maintain its own fleet. Motorized vehicles turned out to be ideal for moving goods through harsh terrain, especially once Pershing's troops were forced to concentrate near the newly built or improved roads along the 100-mile-long route from Columbus to Colonia Dublán. Fortunately, Villistas and Carrancistas alike avoided the area of Colonia Dublán in the wake of Carrizal. Truck companies were able to deliver quantities of materiel that made the men's lives easier than they had been on the trail.

In addition to using trucks to transport materials, the Signal Corps used motorized vehicles as part of its plan to maintain communications in the field. Detachments employed and experimented with radio tractors or truck-mounted radio sets that proved particularly useful in monitoring Mexican communications. The Signal Corps also established two wireless telegraph stations. One remained at Columbus to relay traffic to and from the United States while the other traveled with Pershing's headquarters. In addition to these stations, the signal soldiers laid miles of "buzzer" lines, a battery-powered telephone system that could transmit Morse code over poorly insulated or even bare wires where an ordinary telegraph often failed. Unfortunately, soldiers soon discovered that bare wire laid on the ground as field lines proved vulnerable to wet weather and human and animal traffic. In a state that lacked telegraph lines apart from those that followed the few railroads, the signal soldiers eventually built a permanent pole line from Columbus to Colonia Dublán and worked to facilitate communications between Pershing's headquarters and far-flung detachments. The Signal Corps also installed telephone lines along the entire length of the border where none previously existed. By the end of the campaign, the soldiers had constructed 642 miles of telephone and 677 miles of buzzer and telegraph lines, as well as operated 19 radio stations.

While Pershing relied on trucks to move material and men on the ground during the Punitive Expedition, airplanes were used mainly for reconnaissance, observation, and communication missions. Capt. Benjamin D. Foulois was the commander of the 1st Aero Squadron, which was comprised of eleven officers, eighty-two enlisted men, and one civilian technician. At the start of the expedition, these soldiers had a mere eight JN–3 "Jenny" biplanes made of fabric, wood, and metal at their disposal. The Jennies represented a modest technological improvement over the planes that the squadron used prior to September 1915, the JN–2s, which suffered from a host of major design defects. While the older machines remained in service, Foulois quipped, "An airplane that looks like an airplane may be something less." Even though the JN–3s had an improved design, the small number of them placed the Army far behind European armies in both the investment in aviation and in terms of their ability to use planes in the field. From 1908 to 1913, a time during which the U.S. Army spent about a half million dollars on aviation, Germany spent $28 million. On the Western Front, France and Germany already used airplanes in dogfights, reconnaissance, and short-range bombing missions.

The JN–3s may have been better than previous models, but the Signal Corps' planes still suffered from serious defects that made them inadequate for use in Chihuahua. Perhaps the most serious problem was that the planes were only able to climb to 10,000 feet. Given that the peaks in Chihuahua extend to about 10,000 feet, the underpowered and unstable planes were unable to fight the strong air current flowing over the Sierra Madre. On top of the planes' lack of horsepower and their proclivity for going into spins, pilots were required to fly with field equipment and thirty-four pounds of gasoline. All of this material weighed down the airplanes and made it even harder for them to climb without stalling. The 1st Aero Squadron also had to contend with a serious shortage of replacement parts for its airplanes. Anticipating this problem, Foulois sent a request for parts in March 1916 to Chief Signal Officer Brig. Gen. George P. Scriven who allocated $19,000 to fill the immediate need.

When ordered to support the expedition, the men of the 1st Aero Squadron disassembled their planes, packed their equipment at their base in Fort Sam Houston, and departed San Antonio by train. They arrived at Columbus on 15 March and spent several days reassembling their vehicles and aircraft and flying short test flights. One twenty-mile sortie into Mexico marked the first aerial reconnaissance done by a United States aircraft into a foreign territory. On 19 March, Pershing ordered Foulois and his men to fly to Casas Grandes. The squadron took off at 1710 intending to fly the entire 100 miles without stopping at nightfall—a decision that had serious consequences. Only one pilot had ever flown at night, and the only way for all pilots to maintain course was for each to follow the plane in front of him until they sighted the bonfires that were supposed to mark the runway at Pershing's camp.

The plan went awry from the start. One of the eight planes developed engine trouble immediately after takeoff and landed safely on the runway at Columbus. Three planes then became separated from the other four in the air. One of these pilots, 1st Lt. Robert H. Willis, flew past Casas Grandes until his plane began to run out of fuel, forcing him to land near Peterson, the site of a major lumber mill. Leaving his plane, he hiked north along the railroad tracks until he encountered U.S. troops. The second pilot, 1st Lt. Edgar S. Gorrell, also flew past the intended airfield and sighted what he took to be Pershing's bonfires. On his approach, he realized he mistakenly had guided on a forest fire instead. Gorrell landed and managed to hire a local guide who led him to Ascensión. The third pilot, 1st Lt. Herbert A. Dargue, lost visual contact with the others soon after takeoff and flew to Ascensión, where he landed for the night and resumed his flight to Casas Grandes the next morning. The other four pilots, including Foulois, managed to stay together, but they also were forced to land near Ascensión for the night and flew on to Casas Grandes and then to Colonia Dublán in the daylight. Given that it took several days for the three missing pilots to rejoin the rest at Casas Grandes, the squadron's effective strength dropped from eight to five operational aircraft while flying to headquarters.

Undeterred, Pershing put the squadron to work. The first flight the squadron undertook was a reconnaissance mission on 21 March. Capt. Townsend F. Dodd, with Foulois as observer, flew south for twenty-five miles, following the tracks of the Northwestern Railroad toward Cumbre Pass before high winds blowing off the mountains forced their return. The next day another pilot and observer successfully located and delivered a message to Colonel Erwin in the Galeana Valley and returned without incident. On its third day in Mexico, the squadron experienced mixed success, managing to find Colonel Dodd's detachment to deliver messages but failing to find other columns moving south along the railroad. After returning to base, massive dust storms grounded the squadron until 25 March. When Pershing relocated his headquarters to the south, the 1st Aero Squadron shifted to Namiquipa and flew short liaison missions to maintain communications and provide some reconnaissance support to the flying columns.

Even though some of these missions were successful, the limited usefulness of the JN–3s quickly became apparent in Mexico. On 22 March, Foulois requested ten additional planes and the spare parts necessary to repair and maintain the remaining aircraft. Pershing endorsed the request and passed it to the War Department, where it reached Secretary Baker's desk. Baker then asked Funston for advice about the number of planes needed by the expedition, and the general indicated that he disagreed with Foulois. In his opinion, the 1st Aero Squadron only needed four more planes in Mexico. Foulois repeated his request for ten additional planes on 30 March, and this time Congress chose to act,

Lt. C. G. Chapman of the 1st Aero Squadron taking off in his JN–3 on a scouting mission from Casas Grande (Library of Congress)

passing the Urgent Deficiency Act the next day. This emergency measure allocated $500,000 to the Army for the purchase of twenty-four airplanes plus portable machine shops, trucks, cameras, guns, and other weapons and equipment. Only three days after Congress appropriated the funds, an article titled "Risking Lives Ten Times a Day, but Are Not Given Equipment Needed" appeared in the *New York World*. It quoted Lieutenant Dargue describing the Jennies as "suicide busses" and blamed the Army for creating dangerous conditions through a lack of knowledge about aviation and aeronautics. The article created a minor scandal within the War Department, and Secretary Baker demanded to know if any of the 1st Aero Squadron pilots had spoken directly to reporters. At first, all denied that they had, but Lieutenant Gorrell eventually admitted that he was the one who complained to the reporter. Pershing did not reprimand Gorrell.

On 7 April, Pershing ordered the aero squadron to deliver a request to the United States Consul General in Chihuahua city to seek Mexican government assistance in obtaining supplies. With the state capital far from areas patrolled by the cavalry, Foulois took two planes on the mission to ensure at least one copy of the message reached the consulate. One plane flew to the north of the city and one to the south. The pilots would remain to guard the planes after landing while the observers were to deliver the messages. The plane that landed on the north side of the city made the trip without incident, but Captain Foulois' and Lieutenant Dargue's plane attracted a large crowd south of the city. Foulois ordered Dargue to fly to meet the other plane in the north while he walked to Chihuahua city to deliver the message. As Dargue took off, someone in the crowd fired several shots at the Jenny, and Mexican police arrested Foulois on the ground. On his way to the city jail, Foulois managed to tell a U.S. citizen in the crowd to inform the consulate of his arrest. The pilot also demanded that his captors inform the military governor, General Luis Gutiérrez, of his situation. After receiving the message that a U.S. pilot was in custody, Gutiérrez released Foulois and provided him with a detachment of police to guard both planes, which were north of the city. Despite the presence of guards, the U.S. pilots decided to fly the planes to the grounds of the American Smelting and Refining Company to prevent mobs from causing further damage to their aircraft. One plane escaped, but a rock-throwing mob brought down Dargue's and Foulois' plane before the federal police could arrive. For their own safety Dargue and Foulois spent the night at the consulate and flew off the next morning.

The squadron continued to fly reconnaissance and communications missions for the next couple of weeks, but by the end of April, none of the planes remained operational. Six had crashed and were damaged beyond repair while the remaining two were no longer mechanically airworthy. These were cannibalized to salvage usable parts. On 20 April, the War Department ordered the 1st Aero Squadron back to Columbus to receive the first four new planes purchased under the Urgent Deficiency Act. A second shipment arrived on 1 May. Unfortunately, these planes also were deemed unsuitable for service in Mexico because of a number of manufacture and design flaws. Regardless, the men of the squadron spent the summer testing materiel, experimenting with different types of equipment, and determining the best ways to mount cameras and weapons on the planes. Although they ultimately spent little time flying operations in Mexico, the pilots and mechanics of the 1st Aero Squadron were able to apply lessons learned, train personnel, test equipment under field conditions, and study how to best employ aircraft to facilitate communications and to gather intelligence under combat conditions. More importantly, the Army corrected some of the major deficiencies in its aeronautics program, and the War Department convinced Congress to allocate money to improve readiness and efficiency. By the end of 1916, the Army had ordered 346 airplanes, compared to the 13 planes it had at the beginning of the year.

On the Border and the National Defense Act of 1916

General Funston had relatively few men to patrol the 1,700-mile-long border with Mexico prior to the raid on Columbus. Detaching troops for the expedition only made the task more of a challenge, and the absence of a significant number of regulars seemed to invite more violence. After the skirmish at Parral, General Scott convinced Secretary Baker that calling the National Guard to federal service could help defend U.S. citizens and property, if not deter further cross-border raids. Baker and Wilson initially had denied the request, hoping to diffuse tension and minimize the possibility of war. After the Glenn Springs raid on 6 May, Generals Scott and Funston repeated their recommendation to the War Department. Three days later, President Wilson called up the National Guard from Arizona, New Mexico, and Texas—chosen for their proximity, states' interests, and men's acclimatization to the weather of the region—to federal service to guard the border.

Deficiencies in the National Guard from the three states soon became apparent. By far the most glaring problem was that there were too few guardsmen available for duty. With 984 men from Arizona, 737 from New Mexico, and 2,470 from Texas, the Army had, at best, a total of 4,191 men available to meet the minimum required call-up of 4,000 to 4,500. However, a large number of these guardsmen were physically disqualified for military service. Many units were so understrength the War Department ordered them disbanded. With so many units already well below required peacetime strength levels, each of the three states attempted to recruit more men to fill the ranks, but the task proved difficult given the short time frame. New volunteers also seemed to be confused about the terms of their service, and some believed they had enlisted in the Regular Army. Meanwhile, the National Guard in other states, such as New York, hosted recruiting parties to increase membership in anticipation of being called to the border. Recruiting efforts outpaced the National Guard's ability to build proper facilities, and recruits arrived at training posts like Camp Whitman, outside of Green Haven, New York, only to find barracks and other buildings still under construction.

On the Border, August 1916, *by Donna Neary, depicts a column of soldiers of the 2d Infantry Regiment, Connecticut National Guard, on the march in Arizona.*
(National Guard Heritage Series, Army Art Collection)

Partly in response to the clear weaknesses demonstrated in the National Guard, on 3 June Congress passed the National Defense Act of 1916. The law expanded the National Guard as well as the commitment it required from members. Federal funds were allocated to pay guardsmen for forty-eight drill days per year, double the previous requirement of twenty-four, plus fifteen days of annual training in the summer, an increase of ten days over the previously authorized five. In addition to the augmented training, the law reinforced the National Guard's federal role as a reserve of the Army. In cases of national emergency the Guard could now mobilize and go overseas with the Army until the end of the emergency. Congress also allocated $17 million to purchase more airplanes for the Army, supplementing the Urgent Deficiency Act. Legislators were looking toward the possibility of bolstering the National Guard's force on the border, but they clearly thought ahead to its possible deployment to Europe after a declaration of war.

Soon after the National Defense Act passed, but before its implementation, President Wilson found it necessary to expand the call-up of the National Guard to the border. This was prompted by General Jacinto Treviño's 16 June statement that he would attack U.S. forces if they moved in any other direction except north, which convinced the president that war on the border was imminent. In early June, however, hardly enough guardsmen were in Texas to patrol the border, much less launch a full-scale invasion of Mexico. Wilson responded by calling up the National Guard from all of the remaining states on 18 June and ordering them to assemble on the border. The next day, guardsmen gathered in local armories across the country, and by 1 July, 122 troop trains headed to the border, with another 101 trains to follow the next week. By the end of summer, approximately 110,000 guardsmen were stationed on the border from California to Texas.

Pvt. Robert Batchelder of the 8th Regiment of Infantry, Massachusetts National Guard, a college student, was among these newly mobilized guardsmen. As a youth he had viewed the state militia as "a group of bloodthirsty, hare-brained idiots who drill occasionally and uselessly." He changed his mind when he saw the increased federal funding and interest in the National Guard, and he believed it was his patriotic duty to enlist. Batchelder wanted to contribute to America's military readiness in the event that it entered the war in Europe, but he also entertained visions of marching triumphantly into Mexico City with flags waving. The young Harvard sophomore's dreams of service and glory were reinforced by the celebrations held in Boston after his unit was called into service for duty on the border. The men marched in parade through the city and received gifts on the way to Texas.

Batchelder's dreams were disturbed by the reality of service on the border. Like other guardsmen, he found himself in a difficult foreign landscape in which he could not "properly describe the exquisite discomfort of the place." He likened rocks to "veritable radiators of heat" and found that the rough ground made it "next to impossible to drive pins for the small tents." Because of the heat, the tepid, alkaline water he and his fellow guardsmen drank from their canteens barely quenched their insatiable thirst. Of his daily routine, he wrote, "On account of the monotony of the country, the long hours of patrol seemed endless." The Southwest always appeared to be either too hot or too cold, filled with dust and people living in startling poverty. Batchelder saw comrades faint after drill, unused to performing

The soldiers of Company H, 5th Regiment of Infantry, Maryland National Guard, at Eagle Pass, Texas, were typical of the many National Guard troops. (Library of Congress)

physical tasks in such a dry climate. Particularly for recruits from the Eastern United States—who traded greenery and trees for sand and desert—working in the Southwest required a period of adjustment. Most guardsmen also had to get used to regular physical exertion in the form of six- or seven-mile road marches and daily drills.

In addition to dealing with the often harsh environment, at least initially, the guardsmen had to contend with shortages of supplies and equipment. Batchelder described their arrival at Camp Cotton, Texas, as "filled with disappointment" as he surveyed the stretch of tents within sight of downtown El Paso. He and his comrades slept on the ground in the tents and spent much of their first days digging latrines and drainage ditches to help prepare the camp for the arrival of more National Guard soldiers. Through July, meals mostly consisted of beans due to shortages of rations and cooking utensils. That all changed in August. After a patrol of several weeks, during which they trained on machine guns, dug trenches, and guarded the border, the men of Batchelder's unit returned to find Camp Cotton had been wired for electricity. It now had showers, wood floors in the tents, and a new store where the men could purchase comfort items. The men went from Spartan conditions and a strict diet of beans to enjoying ice cream, tobacco, and toiletries at reasonable prices. For recreation, they could also visit the new YMCA building. The Army also began inoculating the men against typhoid fever, ensuring that even in badly constructed and provisioned camps the men remained relatively healthy. Men in other units experienced similar privations in the first month after their arrival. Members of the 3d New York Artillery, stationed close to Brownsville, reported shortages of

water in July, and equipment necessary for their training did not arrive until August. In early July, General Bliss began a month-long inspection of all of the camps on the border and found guardsmen in a wide range of living conditions. Most of them had encountered some shortages including a lack of weapons and horses in some of the machine gun companies.

Disengagement, Demobilization, and Off to War

With the end of active patrolling beyond the 150-mile cordon in late June 1916, the men of the Punitive Expedition settled into routine camp life. The wind-swept area around Colonia Dublán was prone to dust storms, which made conditions in camp uncomfortable in the first weeks, but soon the troops started constructing more permanent dwellings, mostly small adobe buildings. Men received bedding, field desks, and cooking implements and made make-shift furniture for their adobe huts. Due to the efficiency of the truck convoys, soldiers received sufficient rations and regular mail delivery. The biggest threat facing Pershing's soldiers was the growing boredom of life in a static camp with little action. In order to counter this problem, Pershing instituted a rigorous schedule of training for the infantry, artillery, and cavalry. The men practiced rifle marksmanship, received a special three-month course in machine gunnery, and participated in daily combat exercises. The general was preparing his men to fight in Europe and believed the expedition's troops would form the nucleus of the expanded Army that would soon organize.

During the first days of their restriction to Colonia Dublán and El Valle, the men had little to do except train, but Pershing soon

Soldiers of the 16th Infantry around a campfire at San Gerónimo, Mexico, May 1916 (Library of Congress)

had diversions established to prevent what he called "discontent and despondence." The men constructed a boxing ring and cleared areas for playing baseball and football. Troops and the Apache scouts hunted for game. The men eventually built a small movie theater that showed films nightly, and they staged short plays and minstrel shows. Soldiers were not allowed to have liquor in camp, but local businessmen established strings of cantinas and small stalls that sold food along the edge of Colonia Dublán. On Christmas, a massive blizzard and dust storm coated the camp in a thick layer of sand, but the men washed off their turkeys and game and erected a massive seventy-foot-high tree harvested from a local forest. The Red Cross distributed donated gifts to the men. In addition, Pershing sanctioned the establishment of a "sanitary village" for prostitutes to service the soldiers of the camp. Capt. Julien E. Gaujot of the 11th Cavalry supervised the village and took responsibility for housing the women and ensuring their customers paid them, as well as for aiding them in establishing schedules. As a result, the men suffered a low rate of venereal disease. The Mormons in the area probably disapproved, and some Medical Corps officers urged Pershing to do away with the system of sanctioned and regulated prostitution, but Pershing refused, pointing to the positive results of the arrangement.

As U.S. soldiers settled into the routine of camp life, Villista forces began to recover from their defeats of the past year and a half. By July, Villa had partially recovered from his leg wound and finally agreed to surgery to remove the bone fragments to speed the healing. After fighting several minor skirmishes in early summer, Villista and Carrancista forces again started clashing regularly and with larger forces. On 15 September 1916, Villa slipped into Chihuahua city to reconnoiter the city's defenses and visit friends during the Mexican Independence Day celebrations. At the same time, 600 of his followers infiltrated the area amid the distraction of the traditional reenactment of Father Miguel Hidalgo's *Grito de Dolores*, the speech that called for liberation from Spain in 1810. At 0300 the next morning, 1,000 Villistas launched a surprise attack in the capital. They seized booty and captured the governor's palace, where Villa gave a speech from the balcony. Despite the surprise, General Treviño quickly rallied the Carrancista garrison in Chihuahua city and led a successful counterattack. The recently wounded Treviño declared victory, but Villa enjoyed a moral triumph by demonstrating that he could still muster a large enough force to take and hold the state capital, even if only temporarily. Villa also managed to escape with 1,500 new recruits and captured supplies with which to rebuild his forces.

Villa and his men followed the attack on Chihuahua city with a series of victories at San Andrés, Cusihuiriachic, San Isidro, and Santa Isabel. Government forces suffered from a large number of desertions, shortages of ammunition, and damaged bridges, telegraph lines, and railroads. In response, Carranza replaced Treviño with General Francisco Murguía, but Villa attacked Chihuahua city again in November in anticipation of the arrival of a new governor and of fresh troops from Durango. Meanwhile, Pershing could only watch developments from Colonia Dublán as Villa refrained from provoking U.S. forces into intervening, thereby preserving the delicate balance that existed in northern Mexico between Pershing and Carranza's forces.

Meanwhile, negotiations for the withdrawal of the Punitive Expedition dragged on. The United States formally agreed with Carranza's proposal to create a joint Mexican American commission on 28 July 1916, but the body did not meet until September in New London, Connecticut. Once again, representatives from the two countries could not agree on a timetable for withdrawal. Mexico's representatives maintained that Pershing and his men should leave immediately, while U.S. commissioners insisted they

Soldiers of the 6th and 16th Infantry marching north out of Mexico at the end of the Punitive Expedition, February 1917
(Library of Congress)

would only do so after the Mexican government guaranteed it would keep the border secure. Deadlocked, the commissioners moved the proceedings to Atlantic City, New Jersey, to start a new round of negotiations at the end of September, but the United States persisted in arguing for retention of the right to intervene again in Mexico in the event of future raids. Carranza refused to accept these conditions, fearing that allowing such a right would effectively transform Mexico into a quasi-U.S. protectorate, akin to Cuba after the Spanish-American War. Negotiations continued through November, when the United States began to soften its demand to resolve the stalemate in response to the increasing threat of war with Germany. On 24 December 1916, the commission finally came to an agreement, which called for the expedition to cross the border within forty days. Pershing received the order to break camp in January, and on 5 February his forces crossed back into the United States.

Starting in February 1917, the War Department also began to demobilize the National Guard forces on the border. Patrols of the border continued, but these were conducted by units of regulars. Many Guard units had not completed the movement back to their home stations when Congress declared war on Germany in April. Some National Guard soldiers went almost straight from training on the border to preparation for deployment overseas. For many of the 110,000 men who served in the National Guard from late June 1916 to February 1917, the training they had received helped to prepare them for combat in France and to train the many new recruits who soon joined them.

As the demobilization was taking place, General Funston, who was overseeing the process as the commander of the Southern Department, died suddenly on 19 February 1917 in the lobby of a hotel in San Antonio. His death meant the loss of one of the Army's most experienced and well-known general officers just as preparations were being made for the United States' eventual entry into World War I. Pershing was one of the few officers who approached Funston in experience and public acclaim, and Funston's untimely death made Pershing the most obvious choice to lead American forces into France.

The Punitive Expedition also continued to have diplomatic ramifications for the United States, Mexico, and Germany. On 16 January 1917, the U.S. government became aware that German Foreign Secretary Arthur Zimmerman had sent a telegram to the German ambassador in Mexico with a proposal for Carranza. The so-called Zimmerman Telegram offered that, in return for starting a conflict that would keep the United States occupied on its southern border, Germany would see that Mexico received Texas, New Mexico, and Arizona—territories ceded to the United States in the 1848 Treaty of Guadalupe Hidalgo—after it won the war in Europe. British naval intelligence intercepted and shared the communication with the U.S. Embassy in London. Wilson released the text of the telegram to the public on 28 February, hastening the nation's entry into the war. Although it outraged most Americans, the Zimmerman Telegram may have been easier to dismiss had the United States and Mexico not spent the previous year and a half embroiled in difficult negotiations concerning the violence on the border. The suspected German involvement in the skirmishes at Parral and Carrizal also seemed to reinforce the possibility that the Mexican government might find such an overture tempting.

The withdrawal of Pershing's forces did not close the book on the Columbus raid. In Mexico, Pershing had captured thirteen men accused of participating in the raid. Five of these men died of battle wounds, but the expedition brought the rest back to the United States as prisoners without any formal extradition proceedings. Despite concerns about the ability to receive a fair hearing in a civilian court in New Mexico, the men were tried in Luna County in April and sentenced to death. Although Wilson asked New Mexico's governor to delay their executions while negotiations continued, five men were ultimately hanged and one received a commuted sentence and returned to Mexico. The trial and resulting diplomatic crisis made the Zimmerman Telegram seem more dangerous in that it highlighted and exacerbated the ill feelings that still persisted on both sides of the border. Carranza rejected the Zimmerman proposal, but Germany continued to offer Carranza various incentives to attack the United States throughout World War I, proving that America's southern border still appeared a vulnerability that could be exploited.

Analysis

The Punitive Expedition failed to capture or kill Pancho Villa, and the presence of United States troops in Chihuahua only temporarily disabled the rebel general's army. Some bands of Villistas scattered and were forced to suspend their guerrilla-style attacks in Chihuahua, but when Pershing's forces left, they were able to reunite and operate as before. These groups might have temporarily become inactive even if Pershing had never been in Mexico due to Villa's injury and his withdrawal from active campaigning during the spring and summer of 1916. Pershing's expedition therefore did not succeed in carrying out its mission to permanently disable or destroy Villa's army.

While many factors made finding and capturing Villa a difficult proposition even if he had not been forced to go into hiding, the lack of intelligence on the ground stands out as the key problem that prevented U.S. troops from operating effectively in the field. Pershing's forces simply could not obtain reliable information either from Chihuahuan citizens or from Carrancista forces. Chihuahua was Villa's strongest bastion of support, but Pershing failed to understand that even citizens who refused to support Villa would not necessarily support the United States in its efforts to find the rebel general. Anti-Americanism was rampant throughout Mexico as was resentment of Wilson's apparent determination to intervene in the Mexican Revolution. The War Department was also not equipped to support the intelligence effort on the expedition. Although the expedition did have an intelligence officer, Maj. James Ryan, who attempted to build networks of informants in Chihuahua, his reports went unread and had little to no impact on decision making in Washington.

Both Pershing and Wilson also erred in their belief that the Constitutionalist government would support the presence of U.S. troops in Mexico simply because they had a common enemy. This created a dangerous situation, making clashes between U.S. and Mexican government forces almost inevitable. Full-scale war between the two states remained likely through the summer of 1916. The fact that Pershing's forces assumed they would have the cooperation of Mexican citizens betrayed a poor understanding of the military situation and an attitude of U.S. paternalism toward Chihuahuans. When Carranza's forces refused to support U.S. troops, Pershing and the men of the expedition treated them as cobelligerents with Villa's troops. They wrongly interpreted the unstable military situation in Chihuahua as a failure of the government's forces rather than as being the result of a complex and still ongoing revolutionary civil war.

In addition to these failures, Chihuahua's difficult terrain meant that capturing Villa would have been a challenge even if political circumstances had been more favorable. Chihuahua was remote and sparsely populated, and traveling over long distances required the Army to fundamentally alter the way it moved men and goods. Horses and mules could negotiate wide spaces without the need of roads, but animals proved to be of limited use in places where there was little fodder. Pershing could not establish effective supply lines that could follow the cavalry into remote regions, putting the burden of obtaining necessary goods on the flying columns. In this respect, Villa's men had an advantage in that they knew the terrain well, had home bases that provided some degree of protection from enemies and the elements, and had local sources for obtaining goods. Chihuahua's remoteness and rough terrain also hindered communications between Pershing and the cavalry, making it more difficult for troops to share information, get reinforcements, or obtain provisions when running low.

Despite its failure to achieve its stated mission, the expedition served as an important training ground for soldiers and as a proving ground for new methods and equipment, particularly motorized transport. Regulars gained experience in large, multiunit operations, training for the kind of fighting that they would soon be doing in France even while the United States officially was still neutral. The Army tested equipment and addressed critical deficiencies. At the start of the expedition, the Army had only 105 trucks in service. When it ended, the Army had seventeen truck companies just on the border. As more vehicles became available,

Major Madden, the expedition quartermaster, was able to train Army regulars as drivers and mechanics through the summer and fall of 1916. Based on the experience of the expedition, Madden determined the optimal organization of men and equipment in Army truck companies. Having a wide variety of makes and models in Mexico convinced him and others that the Army should adopt a standardized truck that could meet all of the Army's needs. The Motor Transport Section of the Quartermaster Corps and the Society of Automotive Engineers collaborated to design such a vehicle, which resulted in the development of the Liberty Truck in 1917. Ten weeks after completing the design, the first models became available. Of the 9,500 Liberty Trucks produced, 7,500 went overseas, making the American Expeditionary Forces the most motorized army of World War I.

The Army's air service also gained much knowledge and experience from its employment with the expedition. It transformed from a single unit with 8 inadequate and obsolete aircraft in March to having orders in place for 346 planes by the end of 1916. The communications and reconnaissance missions that Captain Foulois and the men of the 1st Aero Squadron flew in Mexico allowed them to experiment and recommend improvements to Army aeronautics. Applying additional lessons gleaned from the combatant nations of Europe when the flying units of the American Expeditionary Forces went to France, U.S. aviators operated more powerful and sturdier aircraft than they would have imagined possible during the early days of the Mexican expedition.

Wilson's decision to call up the National Guard to the border also led to quick improvements in the force that increased its readiness just months before the United States' entry into the European war. Chief among these was the massive effort to recruit new guardsmen after the call of Arizona's, New Mexico's, and Texas' Guards failed to produce the numbers of men necessary to patrol the border effectively. This led young men across the country to enlist in large numbers. With all of these new recruits and the excitement around the possibility that they might be called to serve abroad, the Guard received a needed influx of young men eager to find ways to help enhance the nation's military preparedness without enlisting in the Regular Army. In the short term, sending 110,000 men to the border for up to nine months meant the Guard could give its new recruits intensive amounts of training. These men also served alongside men from other parts of the nation, allowing them to shake the charge that guardsmen were too locally focused and parochial to fight in coordinated groups with units from around the country before April 1917. By the time these same men were called to France, they were seasoned troops with some experience in military discipline. More significant, they could train others in various aspects of military life.

Perhaps more important from the vantage of the War Department, the Punitive Expedition served to prove Pershing's worth as a field commander. Pershing had experience as a military governor in the Philippines, but his time in Mexico showed that he could command a large force in foreign territory for an extended period of time. Pershing could sometimes still be out of sync with Wilson and Baker—who actively tried to restrict the scope of the expedition and prevent a larger war at the same time that Pershing pushed to conquer Chihuahua—but he seemed to adapt well to working under confining conditions. Pershing also had the loyalty of the men who served with him in Mexico and elsewhere. He seemed to be a commander who attracted talent and some degree of innovation at a time when the Army was becoming a modern professional organization. The Punitive Expedition thus proved to be a training ground not just for the Regular Army and the National Guard but for the leaders who would soon command men in the trenches of the Western Front.

THE AUTHOR

Julie Irene Prieto received her Ph.D. from Stanford University in 2013 and joined the U.S. Army Center of Military History in October 2014 as a Presidential Management Fellow. At the Center, she specializes in the history of the U.S. Army in Latin America. Her most recent article, "The Sword and the Book: The Benjamin Franklin Library and U.S.-Mexican Relations, 1936–1962," was published in the December 2013 issue of the journal *Book History*.

FURTHER READINGS

Clendenen, Clarence C. *Blood on the Border: The United States Army and the Mexican Irregulars*. New York: Macmillan, 1969.

Eisenhower, John S. D. *Intervention! The United States and the Mexican Revolution, 1913–1917.* New York: W. W. Norton, 1993.

Gilderhus, Mark T. *Pan American Visions: Woodrow Wilson in the Western Hemisphere, 1913–1921.* Tucson: University of Arizona Press, 1986.

Harris, Charles H., III, and Louis R. Sadler. *The Great Call-Up: The Guard, the Border, and the Mexican Revolution.* Norman: University of Oklahoma Press, 2015.

Hurst, James W. *Pancho Villa and Black Jack Pershing: The Punitive Expedition in Mexico.* Westport, Conn.: Praeger, 2008.

Katz, Friedrich. *The Life and Times of Pancho Villa.* Stanford, Calif.: Stanford University Press, 1998.

Miller, Roger G. *A Preliminary to War: The 1st Aero Squadron and the Mexican Punitive Expedition of 1916.* Washington, D.C.: Air Force History and Museums Program, 2003.

Pershing, John J. *My Life before the World War, 1860–1917: A Memoir.* Edited by John T. Greenwood. Lexington: University Press of Kentucky, 2013.

Stout, Joseph A., Jr. *Border Conflict: Villistas, Carrancistas and the Punitive Expedition, 1915–1920.* Fort Worth: Texas Christian University Press, 1999.

Welsome, Eileen. *The General and the Jaguar: Pershing's Hunt for Pancho Villa: A True Story of Revolution and Revenge.* New York: Little, Brown and Co., 2006.

This booklet is in the U.S. Army Campaigns of World War I series published by the U.S. Army Center of Military History. For more information on this and other commemorative pamphlets, please visit the U.S. Army Center of Military History Web site at www.history.army.mil.

Made in the USA
Monee, IL
13 September 2026

62981869R00046